CAMBRIDGE COUNTY GEOGRAPHIES

General Editor: F. H. H. Guillemard, M.A., M.D.

ISLE OF WIGHT

Cambridge County Geographies

ISLE OF WIGHT

By

TELFORD VARLEY

Head Master, Peter Symonds' School, Winchester

With Maps, Diagrams, and Illustrations

CAMBRIDGE

AT THE UNIVERSITY PRESS

1924

CAMBRIDGE UNIVERSITY PRESS
Cambridge, New York, Melbourne, Madrid, Cape Town,
Singapore, São Paulo, Delhi, Mexico City

Cambridge University Press
The Edinburgh Building, Cambridge CB2 8RU, UK

Published in the United States of America by Cambridge University Press, New York

www.cambridge.org
Information on this title: www.cambridge.org/9781107628700

First published 1924
First paperback edition 2013

A catalogue record for this publication is available from the British Library

ISBN 978-1-107-62870-0 Paperback

PREFACE

THE author gratefully acknowledges his indebtedness to the many friends who have given him valuable assistance towards the following pages. To the late Rev. Canon J. Vaughan for reading the MS. dealing with Island Flora, to the Rev. J. E. Kelsall for revising the proof-sheets treating of the fauna, and for the interesting details concerning the ravens and peregrines of Freshwater Cliffs: to Dr Williams Freeman for reading the proof-sheets and for a number of valuable suggestions: to Messrs Samuel White and Co. for information as to maritime engineering: to Mr A. F. Rice of Great Park, Carisbrooke, and Mr L. L. Jennings, of Winchester, for assistance in the chapter on Agriculture: to Mr L. A. Cozens, County Highway Surveyor, and to Mr W. E. Madams, of Winchester, for information concerning the highways and official administration, and to a number of others, who have kindly afforded aid.

The maps on pp. 17 and 60 (published in the Geological Survey *Memoirs*) are inserted by permission of the Controller of H.M. Stationery Office. The Geological sections (pp. 26 and 30) are adapted from the publications of the Geological Survey. For certain details on climate on p. 62 the author has consulted Mr John Dover's article on Meteorology in Morey's *Natural History of the Isle of Wight*.

April, 1924

CONTENTS

ILLUSTRATIONS

The illustrations on pp. 7, 11, 12, 14, 19, 44, 47, 48, 49, 51, 56, 64, 72, 73, 75, 79, 80, 98, 100, 102, 107, 108, 109, 110 (both), 111, 125, 126, 127, 128, 129, 130 are from the author's own photographs; those on pp. 33, 57, 88, 94, 105 from photographs by F. Frith & Co. Ltd.; those on pp. 45, 67, 82, 101 from photographs by Valentine & Son, Ltd.; those on pp. 71 and 78 from photographs supplied by Mr A. Rice and Messrs Samuel White & Co. respectively.

1. County and Shire.

England is divided into areas curiously dissimilar in size and shape, known as "shires" or "counties." This division carries us back to the days of the Anglo-Saxon occupation, the period of the so-called Heptarchy.

As the Anglo-Saxons in the 5th century A.D. began to spread over the country, the land became increasingly covered with small communities, till ultimately the whole area was carved out into a number of separate properties, practically independent one of another, and known for the most part as "tuns," "worths," or "hams." Extended occupation in combination with tribal rivalries and relationships tended steadily to bring about greater cohesion, in this way kingdoms more or less stable were evolved, each kingdom becoming later divided into smaller areas called "scires," the modern shire. A shire is a share, *i.e.* a portion shorn off from something larger. It is from this "shire" organisation, with its administrative officers the Earldorman and Scire gerefa, *i.e.* shire reeve (cf. our modern alderman and sheriff), its "tunships" grouped into "hundreds," each with its own administrative assembly or "moot," that all our subsequent developments of local government have grown. From the early "tun" grew the "demesne" or manor of later Anglo-Saxon and Norman days, and from this under ecclesiastical influence the parish ultimately developed.

The word "county"—the practical equivalent of "shire"—has a different origin, being derived from the Norman title of *comte* or count—the *comes* (companion) of the king.

The county is therefore the "comitatus" or area governed by a count. "Shire" tells us of early Anglo-Saxon development; "county" recalls the subjugation of the Saxon to the Norman invader.

The word "county," as used nowadays, is however an administrative term difficult of concise definition. It really means a district set apart for purposes of local government, but as different grouping has to be adopted for different purposes, *e.g.* administration of justice, Parliamentary representation, home defence, education, etc., the precise meaning of the word "county" varies according to circumstances. For instance, an area, *e.g.* a municipal borough, may be independent as regards certain activities, and yet be included in a county area for others, so that while we still speak of the old geographical county in its widest sense as "the ancient county," we have now as the result of the passing of the Local Government Act in 1888, a series of what are called "administrative counties," *i.e.* areas smaller than the ancient county, each governed by its own administrative county council.

In this respect the position of the Isle of Wight is quite *sui generis*, for before 1293, when the Crown repossessed itself of its direct right over the island, it was in reality an independent unit, owing service only to the Feudal Lord, whose position was so independent that the Crown at times regarded him as a menace to its own authority. From 1293 onwards it tended to become merged by natural process more and more into the "ancient county" of Hampshire, until in 1888, under the Act referred to above, it was constituted an administrative county. Accordingly, while the island has now become a "county," it has never been a "shire."

The derivation of the name Wight is uncertain. Diodorus Siculus, in a much-quoted passage, speaks of an island "Ictis," to which tin was conveyed from the mainland by carts "at low tide, all being dry between it and the island," but whether by Ictis is meant the Isle of Wight or St Michael's Mount in Cornwall is disputed. All seem agreed, however, that the word "Wight" is of Celtic origin, possibly connected with a Celtic word *gwyth, a channel.* The Romans called it Vectis or Vecta, early Saxon writers speak of it as Wiht, and the inhabitants as Wihtgara, the Roman and Saxon names being probably both merely adaptations of the name given to it by the ancient Britons. The characteristic popular name for it, both locally and on the adjacent mainland, is "The Island"—while in official language "The Isle of Wight" is the style invariably employed.

2. General Characteristics.

The Isle of Wight lies midway athwart the Hampshire coast and separated from it by the two sea channels, Spithead[1] and the Solent, varying from $\frac{3}{4}$ to 3 miles or more in width. Its surface, soil, and climate, give it a character of its own. It is often spoken of as the Garden Isle. The beauty and variety of its scenery, its sunny and genial climate, together with coast attractions of quite special type, have served for generations to attract an annually increasing number of summer visitors, and the residential population shows a no less steady gain.

Apart from the occupations connected with visitors, and shipbuilding, the industry of the island is almost

[1] See foot of p. 6.

exclusively agricultural, but other industries are now springing up and its ports tend to become more and more centres of industrial activity.

As regards surface the island falls into two natural divisions. From west to east a ridge of chalk stretches in a practically unbroken line from the Needles to Culver Cliff. The portion lying north of this is generally speaking flat, rather featureless, and covered with considerable stretches of woodland, chiefly oak. It is well watered and has numerous ponds of small size.

South of this chalk backbone the island shows essentially distinctive features. Rolling chalk downs, often with stone beacons or obelisks on the chief vantage points, extend more or less radially southward, varied by stretches of sandy and clayey soil, for the most part sloping inward, so that this southern portion is roughly saucer-shaped, and consequently is spoken of locally as "The Bowl." Throughout the whole of this region there is but little woodland. The views over it and over the adjacent sea from many parts of the upper levels are very fine. From such commanding points as the top of St Catherine's Down, Boniface Down, or Carisbrooke Castle the eye can wander uninterruptedly over miles of varied landscape, coast or mainland.

But attractive as the island is on the landward side—and to those who know it best its most appealing parts are inland—it is to the seaboard that the attention of the visitor is usually directed. There is so much of variety and interest in form and colour, as well as in the geological succession of rocks exposed, so much to be seen here which can be seen practically nowhere else in England. Especially notable is the greenness of the land, the verdure reaching almost to the water's edge, a feature it displays chiefly along

Castle and Village of Carisbrooke

(From an old Engraving)

the northern coast, round Cowes and by Osborne. Elsewhere the scenery changes entirely in type, from a flat and featureless coastline along the northern shore to striking chalk cliffs three or four hundred feet high as at The Needles and Culver Cliff, while both the western and eastern faces of the island present quite distinctive features.

One special region, described in fuller detail later, is the coastal fringe along the south from St Catherine's Point past Ventnor to Bonchurch, known as the Undercliff. Here, at a period geologically recent but historically remote, a great landslide has occurred, forming a continuous terrace or false cliff-top on a lower level, sheltered from the north by the towering wall of the original rocks from which it separated. Along this belt the general conditions of climate are those of the warm temperate zone. It is therefore in high favour as a health and winter resort.

Strategically the island forms an effective shield to the vital centre of British naval activity. Commercially it has but little importance except indirectly. The sea channels between it and the mainland possess special tidal advantages and a sheltered character which they owe entirely to the presence of the island. Were it not so one might find some part of the island serving as an advanced post for foreign trade, as, *e.g.* Copenhagen in the Danish archipelago; but the special advantages with which its situation endows these inland waters preclude this, and the island remains in the line of commerce but not of it.

The term Spithead really denotes the head of the Spit off Portsmouth, where men-of-war anchored. To apply it, as is the practice nowadays, to the maritime channel adjacent, although convenient, is, strictly speaking, inaccurate.

Freshwater Bay

3. Size. Shape. Etc.

The island has a total area of 94,146 acres (of which 225 acres are water), practically one-tenth of the acreage of the mainland of the county of Hampshire, of which it is geographically merely a portion which has become detached. It is therefore a very small unit when compared with some of the ancient or geographical counties on the mainland. Yorkshire, for instance, is about forty times as large, but on the other hand, it is only a very little smaller than Rutland, the smallest English shire. There are three Scottish geographical counties, viz. Linlithgow, Kinross, and Clackmannan, which are actually less than the Island, but no Welsh or Irish county anything like as small.

In shape, as well as in orientation, the island shows a curious regularity and symmetry. It is practically a geometrical rhombus, or lozenge, in shape, with its diagonals running, the one due east and west, the other due north and south. The longer axis, from the Foreland to the Needles, is 22½ miles; and the shorter, from Cowes to Rocken End, 13 miles in length. This symmetry is borne out by its position, its northern apex being almost directly opposite the mouth of Southampton Water, and the opposite shores of the mainland running almost parallel to its own coastline. Cliff End, not quite a mile from Hurst Spit, is the point nearest to the mainland.

The nature of the coastline, moreover, carries out the same idea of balance and parallelism. From apex to apex on the southern side, the coast is remarkably regular and free from indentation. On both the eastern and western sides are two broadly sweeping or curving bays—or rather bights—Sandown Bay and what we shall term the Western Bight, for locally the latter has no general name.

In contrast with this, the northern coast, though following a generally continuous line, is cut into by a number of shallow and winding creeks or estuaries, but these, though tidal, are riverine in character, so that while the bolder southern coast of the island nowhere presents the baffling irregularities of the Scottish or New Zealand fjords, neither are the low-lying shores of the north honeycombed like the shores of Holland or Denmark. There are, moreover, no subsidiary islands.

4. Surface and General Features.

While on the one hand the island is actually a detached portion of the mainland, it is in its structure in no way a mere continuation of mainland Hampshire. Indeed, in most respects it presents a decided contrast. Its chief outstanding feature is the elevated chalk ridge, which runs in an unbroken line from west to east, extending from The Needles to the Foreland and Culver Cliff. It keeps a high level almost throughout, and there are only a few places where it falls below 100 feet in elevation. On Freshwater Down it reaches 463 feet, on Mottistone Down 667, at Brighstone Down 701. At Freshwater Gate and Yaverland it falls almost to sea level, but rises again at the eastward end, forming the broad swelling Bembridge Down, 343 feet high, and the bold headland of Culver Cliff.

The breadth of this ridge varies considerably from point to point. Near the eastern and western ends it is quite narrow, but between Brook and Carisbrooke it expands into a series of broad downs, dropping here and there to lower levels, which serve as "passes" for some of the roads leading to the coast from the interior. The geographical as well as geometrical centre of the island is at Newport, and from here radiate a series of valleys sloping gently to the north and draining the central part of the island "Bowl," the upper Medina and its tributary streamlets all uniting at Newport itself to form the tidal estuary or Medina proper.

Southward the elevation again increases, and all along the south runs a great bastion of greensand, capped at its highest point by chalk, forming a series of broad downs, rising here and there to nearly 800 feet. This southern bastion or plateau is intersected by three deep clefts which give access to the interior. The presence of the chalk on the upper levels, both here and along the central ridge, is marked by the line of chalk quarries, from which chalk is dug both for building material and also for "marling" the land.

But what chiefly differentiates the island from mainland Hampshire is that while on the mainland the water-bearing strata (chalk and upper greensand mainly) form the core or fundamental strata of the county, resulting in the collection of the drainage into well-defined streams, in the island the reverse is the case. Here the fundamental stratum is composed of clay, and the chalk and upper greensand form merely a capping over it. Hence the insignificant character of island streams, which otherwise would be more in evidence and certainly better defined.

Moreover, the rocks which underlie the chalk and upper greensand—a series of clays mingled with sands—are very readily worn away by water and atmospheric agencies, so that the valleys tend to become deep swampy and waterlogged hollows in which peat accumulates; and though within the last century much of this peat has been removed and extensive artificial drainage carried out, these natural conditions still persist.

It is to the existence of these deposits of clay as the foundation strata of the island that the special nature of much of its surface features is really due. Water soaks down through chalk or sand till it meets the clay beneath,

Landslip phenomena, Unploughed Balk, St Catherine's Down

which being impervious, arrests its descent, with the result that the upper surface of the clay becomes "unctuous" as it is technically termed, and, where the strata show any appreciable tilt, forms a slippery inclined plane over which the superincumbent masses tend to slide downwards. It was some such process which ages ago produced the Undercliff; but over wide stretches inland, as well as round the coast, the surface reflects this unstable condition, showing big hummocks or billowy folds, which may often be traced for long distances from the centre of active disturbance. Much of the arable land accordingly presents a curious wavy and irregular surface here and there, too rough or steeply inclined for the plough to smooth down, leaving "balks," or islands as it were, of unploughed land, dotted over the general cultivated surface. The beds of clay which are the cause of these phenomena are spoken of locally as

Blackgang Chine

Blue Slipper, the clay being mostly blue in colour, and the term "Slipper" being applied in the island to any band of clay which may cause land to slide, quite independent of its geological formation. This tendency to slide is further accentuated owing to the dip of the strata, which on both the northern and southern coasts slope downwards to the sea, but while on the northern coast it has formed no such striking feature as the Undercliff, it is in places much more continuously in evidence, and considerable stretches along the north are for this reason highly unstable.

On the eastern and western coasts we find the curious "Chines." These are narrow, deep and winding clefts, sometimes of great length, and almost always with precipitous sides, and are quite one of the most picturesque features of the island. Their presence is due mainly to the sandy rock flanking the eastern and western coasts, most of which is very easily worn away by running water. To the east we have Shanklin Chine and Luccombe Chine—on the west Blackgang Chine, and a whole series of less striking but still noteworthy chines. Shanklin and Blackgang Chines have both been artificially "improved," but the latter still remains one of the grandest natural coast features of the island. The whole series of chines, great and small, deserves to be studied. Fortunately all are accessible and apart from Blackgang and Shanklin are entirely untouched and unspoiled.

The southern upland is almost all under cultivation. Most of it is arable, as the mixture of clayey and sandy elements in the soil affords good cornland, and but little of this part of the island carries timber. The highest levels of all are grass-covered downland or waste, and here and there are small patches of wood or belts of trees. These belts of trees are not fortuitous in occurrence, but depend on the presence of strong soil beneath. The waste lands are covered with gorse and bramble, or sometimes, though less frequently, with heather.

The wind-swept nature of certain parts, more particularly near the coast, is shown by the boughs and trunks of trees and bushes, which, owing to the prevalence of westerly winds, develop a pronounced list or inclination. Some of the chines on the western side are filled along their lower levels with a luxuriant growth of timber and underwood, but as soon as this growth reaches the upper levels where it is no longer sheltered, it ceases, and the branches become stunted. The gorse and bush plants, where exposed, grow out to leeward, *i.e.* to the east, until they develop an almost continuous flat matted surface.

Trees bending towards the East, "The Eastward position," Brook

A peculiarity worthy of remark is afforded by both the eastern and western extremities of the island. Both terminate in elevated masses, which jut out boldly into the sea, but in each case they are separated from the rest of the island by well-defined gaps of very low elevation, viz. Freshwater Gate and Sandown Flat, so low that a very few feet of depression or even a specially high tide would suffice to cut them off from the island. Frequently in winter the tidal and wind conditions are such as to cause anxiety on this score. The Western Yar under conditions of specially high tide swells up its estuary with such violence as to threaten to carry away Freshwater Gate, while Sandown Flat, which is practically at sea level, would very possibly have been engulfed in a similar way before now, but for the strong defensive sea wall erected to the east of Sandown.

5. Water Supply. Drainage. Rivers.

The northern and southern portions of the island present quite different problems in water supply. North of the central ridge the London Clay, which here overlies the chalk, forms the floor. It has a slight inclination downwards to the north and is topped by a considerable depth of gravelly and sandy deposits. Hence there is an abundant water supply and a number of streams feed the numerous estuaries of the northern coast. Many wells have been sunk in this area, *e.g.* at Bembridge, Freshwater, Haven Street, Newport, Parkhurst, Ryde, West Cowes, etc. Wells in Bembridge have been sunk to over 700 feet in depth.

South of the London Clay the disposition of the water-bearing strata is much more complex. The chalk is the main source of supply. Freshwater, Yarmouth, Newport, and Ryde waterworks all derive their water from wells sunk in the chalk of the median ridge, Ventnor from springs in the chalk of Boniface, Chale from springs at a high level on St Catherine's Down. Shanklin, however, taps the upper greensand. But, in spite of abundance of water, the disposition of the rocks tends to disperse the drainage and not to collect it. All along the southern face the southward dip of the strata gives the drainage a direction seawards, producing thereby the conditions of landslide which ages ago formed the Undercliff, so that instead of contributing to the island catchment or riverine system its effects are mainly destructive of the stability of the cliff face. These southern strata in fact are disposed for riverine catchment purposes as ineffectively as an inverted soup-plate in a shower of rain. The slightly concave centre would retain a certain amount of water, but elsewhere it would flow off down the slopes on every side.

But while, apart from the Medina estuary, the actual rivers are insignificant, their distribution is interesting and instructive, and the study of their basins takes us back to those very early days when southern England was emerging as dry land from the sea. The whole emergent surface formed at that time a series of great folds, with axes running roughly east and west over England, and with troughs between each crest. From the earliest period of emergence, however, drainage commenced, and with it erosion—channels began to be worn out and streams to flow down the lines of steepest gradient. When ultimately the troughs themselves emerged as dry land these formed the channels which collected the drainage and conducted it to the sea. Such a trough existed along the line now occupied by the Solent and Spithead, and along this an important river flowed, which nowadays we speak of as the Ancient River Solent, the mouth of which formed a wide estuary somewhere near present-day Brighton. The island streams existing to-day, as may be seen on the accompanying map, are all of them remains of tributaries which at that period drained into this Ancient River.

Looking at surface conditions as they exist to-day we should expect to find the median chalk ridge the watershed

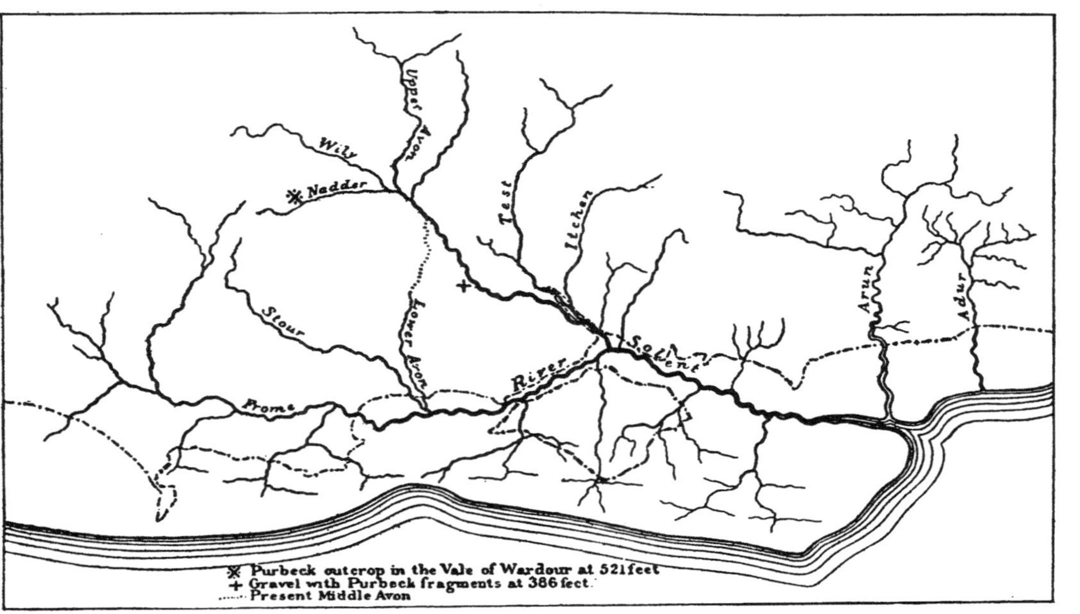

Map of Ancient River Solent showing its relation to present coast outline (· — · — ·)

of the island, but we must remember that it is the lower levels where drainage collects, and not the upper levels where it starts, which determine the ultimate course of a river. Not only was the Hampshire "continent" originally much more elevated than at present, but it extended considerably further to sea—eastward and westward as well as southward—and the island drainage from the very commencement flowed northward to the big Solent trough. Thus streams of much greater volume and therefore of far greater erosive power than those of to-day were actively engaged in cutting out deep river valleys, and carrying away detritus from all the upper levels. Accordingly, while the surface contours have all been literally transformed, the river beds of the ancient island rivers have never materially altered since they first began to flow. In this transformation active coast erosion has also played its part. Not only has our area now become an island, but much of its eastern and western extremities have been shorn off. Thus most of the actual course of the Western Yar, as well as a considerable part of the Eastern Yar, have disappeared, and some of the springs which now work downward on the southern face of the Undercliff probably once helped to swell the waters of the Medina.

Most of the drainage of the Bowl of the island finds its way either to the Medina or the Eastern Yar. The Medina commences from springs in the upper greensand near Chale Green. Following a slightly winding course northwards, it receives at Shide the waters of the Blackwater—a name reflecting the peaty bottom of the latter—and other streams, the chief of which is the Lukely, which drains the Carisbrooke valley and joins it at Newport. At this point the channel broadens into a tidal estuary, and from here to the

Carisbrooke Village and Valley of the Lukely

sea—some four miles—the Medina is a considerable waterway, navigable at high tide for vessels of coasting type, but shrinking at low tide to quite narrow dimensions.

The Medina estuary is the deciding factor in the commerce and industry of the island, and this combined with the increasing growth of Southampton and the naval activity of Portsmouth mark it out for great and probably rapid development in the future. Southampton, Portsmouth, and Cowes indeed seem mapped out as a natural triangle for the development of marine industries, much as are Glasgow, Greenock, and Dumbarton on the Clyde, and New York, Brooklyn, and Jersey City on the Hudson.

The Eastern Yar presents a more definite riverine character than the Medina. Commencing from streams in the greensand by Wroxall, Whitwell, and Godshill, it winds about eastward past Newchurch, with its banks on either side much undercut and lined with undergrowth, till it reaches Alverstone Mill. At this point the valley broadens out into a wide alluvial flat, known as Sandown Flat, one of the best grazing spots in the island, and originally part of a much larger expanse.

Up to about forty years ago the last two miles of its valley—then known as Brading Harbour—used to be covered at high tide, but in 1880 a permanent embankment was erected across the mouth, and a large area—very nearly one square mile—was reclaimed, and now forms valuable grazing land. Sir Hugh Middleton—of New River fame—tried to reclaim it in James I's reign and lost some £7000 —a fortune at that period—in the enterprise, as a furious high tide swept away the embankment before its completion. An ancient well, considered to be of the Roman period, was discovered in the reclaimed area, showing that subsidence had since lowered the general level.

The Western Yar is an even more insignificant fragment of the river of earlier days, as is shown by the occurrence of river gravels which persist as a capping on the sea-cliff near Freshwater, the bed of the stream and the area it drained having been swept into the sea. Some of the trickles down the western chines, *e.g.* Grange Chine, were once feeders of this Western Yar. Its present source is a small spring at Freshwater Gate, known as the Rise of Yar. Though yielding fresh water, it shows the peculiarity of rising and falling with the flow and ebb of the tide.

The deep wells sunk at the forts at Spithead into the sandy strata under the sea bottom are worthy of remark. At Horse Sand Fort the sandbank is 28 feet below high-water mark and the depth of the well 570 feet. At No Man's Fort the sandbank is 37 feet and the well 579 feet and water rises in the well to within a foot of high-water mark. As the strata on both sides of Spithead are continuous, the question arises whether the water derived from deep-seated wells on one side may not be in part derived from a catchment area on the other. It would be interesting if the island water supply could be shown in this way to be in part derived from the mainland or *vice versa*. The continuity of the water-bearing strata both above and below the London Clay shows this to be by no means improbable.

6. Geology.

The rocks of which the Isle of Wight is composed are all "sedimentary," *i.e.* they have been deposited in layers at the bottom of a sea or other expanse of water.

At the time when the island of to-day began to be formed the east and south of England area was the bed of a shallow sea whose shore stretched north-east from Dorset over the Midlands. Our region at that time was the bed either of the estuary of a great river flowing eastward, or of large lakes or lagoons near the seashore. The climate was warm, and quaint animal forms abounded. Great reptiles, often winged, lived in the swamps, and giant mollusca and crustacea in the warm seas. Over the bottom of these waters, during the Wealden period, sandy and clayey detritus began to settle, in much the same way as river deltas are formed to-day, and so the earliest of Isle of Wight formations came into existence—the Wealden. Numerous reptilian remains, much drift timber, and other parts of plants characterize these strata, a clear indication of deposition in shallow water not far from a shore. The Wealden period was followed by a general subsidence, and a large sea—the so-called Neocomian Sea—was formed, under which the greater part of England and much of Europe eventually sank. On the bed of this sea a great thickness of alternating sands and clays—the results of erosion of adjacent land surfaces—was deposited. Thus we obtained the Lower Greensand, Gault, and Upper Greensand. They form the surface of much of the island of to-day and the foundation of the remainder. Then, as the depth increased, the products of land erosion ceased to be deposited and material almost entirely of marine origin, mainly the hard calcareous parts of minute creatures known as foraminifera, began to collect at the bottom. Thus the chalk was formed—at first impure and without flint nodules, and later hard, white, and abounding in flints, and classified according to its age of deposition as Lower, Middle, and Upper Chalk respectively. The flint, so characteristic of the Upper Chalk, was formed by the deposition of silica dissolved in the sea-water, and sponges which then grew in great quantity on the sea floor served frequently as nuclei for its deposition.

Then the stages of the process were reversed—the sea floor again rose—and the greater part of Britain became a land surface once more and so remained for some time. Then another reversal; the south-east of England sank, and on it great thicknesses of sand and clay were deposited. These rocks we term the Eocene. In the island they begin with mottled red and purple clays (the Reading beds), succeeded by a dark blue or brown clay about 200 feet in thickness—the well-known London Clay—and then by a succession of beautifully tinted red and yellow bands (the Lower Bagshot sands), the exposure of which at Alum Bay is one of the most attractive features of the island. Above these came a great thickness of other sands and clays, the Bracklesham and Barton Beds and the Headon Hill Sands. During the Eocene period the climate was tropical and big mammalia, such as the Palaeotherium, had come into existence.

The next period is one of extreme interest in island geology. Lacustrine or estuarine conditions succeeded the more definite marine conditions of the Eocene, and in the shallow waters so formed a series of deposits collected in

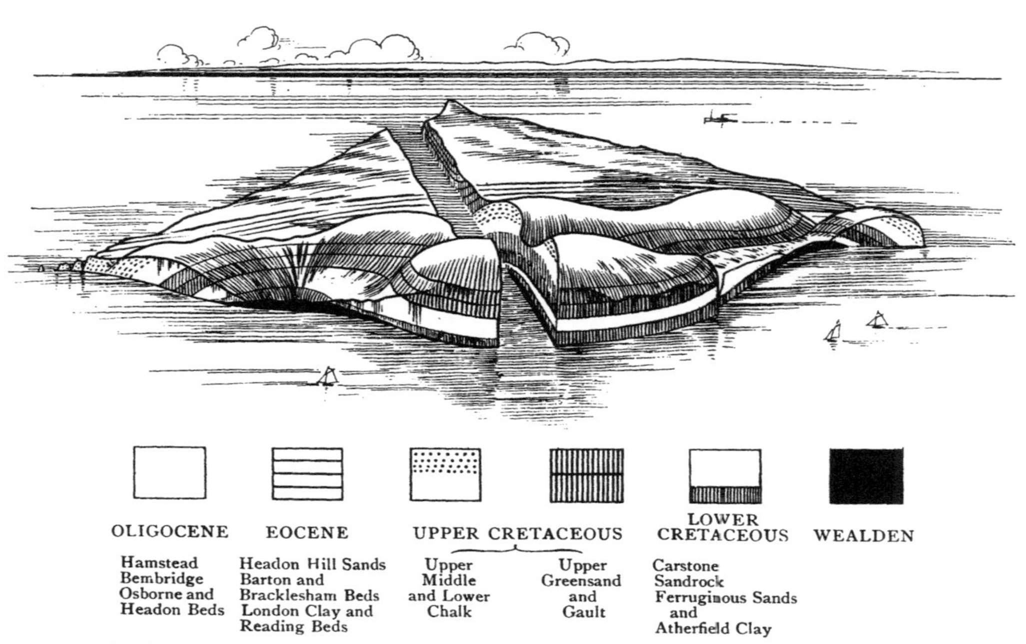

Bird's-eye diagrammatic view of the Geological Structure of the Isle of Wight

the region of the Solent and the contiguous areas. These deposits—clays, sands, marls, and limestones, the Headon, Osborne, Bembridge, and Hamstead Beds—are known as the Oligocene and are represented nowhere else in England.

The climate still continued tropical and the general conditions must have been much as those of the Central African lakes to-day, as remains of palms and other tropical plants, crocodiles, tortoises, and similar creatures are found in abundance.

It was at the close of the Oligocene that the island began to assume its present form. A prolonged movement of contraction and dislocation of the strata of Southern England took place, and under the influence of a great lateral thrust from the south the strata were crushed up into a series of huge folds. Present-day Hampshire and the island began to emerge, at first as ridges running east and west with sea between; the process continuing until eventually the whole south and east of England became dry land again, with our island not as yet detached from the mainland, and a coast-line extending into the English Channel considerably southward of its present limits.

Since this period, though changes of level have occurred more than once, neither the island nor adjacent Hampshire has ever been entirely submerged, and as already related it was at this period that the river system still in existence took its rise. Destructive agencies of various kinds have indeed subsequently removed enormous masses of material. Over much of the area the chalk has entirely disappeared, and the sands and clays underlying it have been not only laid bare, but in many places also removed. Nevertheless, it is as a survival—or, if one prefers it, as a ruin—of the land mass which emerged in post-Oligocene days, and not

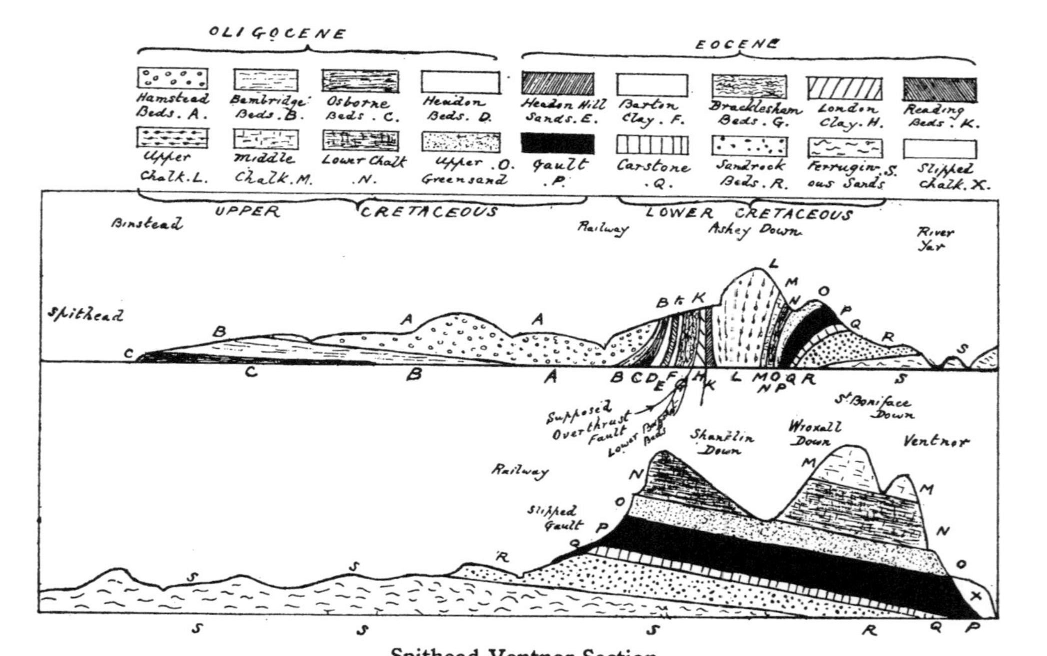

Spithead-Ventnor Section

as a reconstruction of it, that we must regard the Isle of Wight of to-day.

Though there is much complexity of detail here and there, the general structure of the island is really simple to follow, and a few minutes' study of the bird's-eye view on p. 24—adapted from Mantell's well-known book on island geology—will give a better idea of it than would pages of description. If, after studying this, the reader will turn to the Geological section—Spithead to Ventnor here given—he will readily form an idea of the strata represented and their disposition in two great folds once complete and covering both Hampshire and the island, but of which the greater part has since disappeared. He will note the remarkable evidences of lateral pressure under which the Eocene strata and the chalk have in places assumed a practically vertical direction, and he will also realise how enormous has been the denudation process which has removed all the chalk and upper cretaceous beds which were once continuous from the central ridge to the Channel coast.

This denudation was due in large measure to the subglacial conditions prevailing after Oligocene times, which also caused the destruction of the existent flora.

The actual strata of the island, as recorded in the Geological Survey memoir, are as below:

Recent	Blown Sand Alluvium Peat
Pleistocene	River Terraces (gravel) Angular Flint Gravel of the Chalk Downs Plateau Gravel
Oligocene	Hamstead Beds Bembridge Marls Bembridge Limestone Osborne Beds Headon Beds

Eocene	Headon Hill Sands	
	Barton Clay	
	Bracklesham Beds	
	Lower Bagshot Beds	
	London Clay	
	Reading Beds	
Upper Cretaceous	Chalk with Flints	
	Chalk Rock	
	Middle and Lower Chalk with Melbourn Rock	
	Chloritic Marl	
	Upper Greensand	Chert Beds
		Sands
	Gault	
Lower Cretaceous	Lower Greensand or Upper Neocomian	Carstone Sand-rock series
		Ferruginous Sands
		Atherfield Clay
	Wealden Beds with beds of sandstone	

The lowest series are well exposed along the shore of the Western Bight, where an unbroken succession of strata from the Wealden right up to the chalk can be followed in a remarkable anticline stretching from Freshwater to Rocken End. Numerous fossils are obtainable here when winter storms have left them exposed, among them the so-called Atherfield prawn (*Meyeria magna* or *M. vectensis*), occurring in the clay near Atherfield, and a wide series of cephalopods and bivalves, e.g. *Perna Mulleti*, *Gervillia*, *Ostrea*, and ammonites of great size, though the chief fossil prizes are the remains of great saurians—Iguanodon and others. Of extreme interest also is the so-called Pine-raft, a confused mass of fossilised tree-trunks, many of considerable size, lying submerged except at low water just off Brook Point. Another interesting feature along this

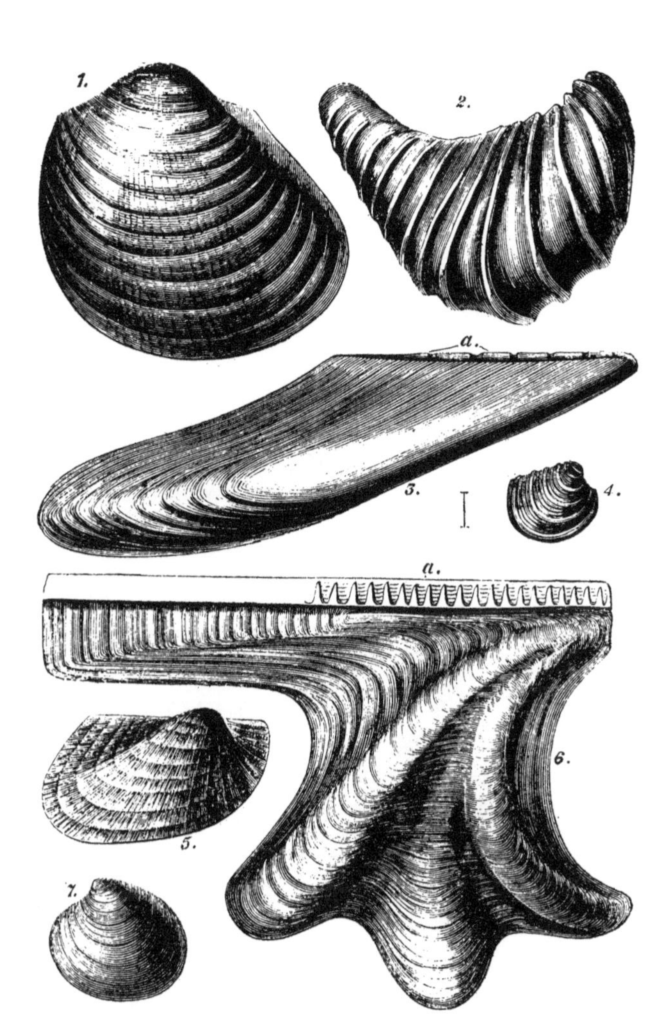

Fossil Shells from the Lower Greensand

1. *Corbis corrugata.* Sand rock, Atherfield. (Half natural size.)
2. *Trigonia caudata.* Sand rock, Atherfield.
3. *Gervillia anceps.* Cracker rocks, Atherfield. (Half natural size.)
4. *Venus striato-costata.* Cracker rocks, Atherfield. (Double natural size.)
5. *Arva kaulini.* Sand rock, Atherfield.
6. *Perna Mulleti. Perna Mulleti* beds, at junction of Wealden and Lower Greensand. (Half natural size.)
7. *Venus parva.* Shanklin.

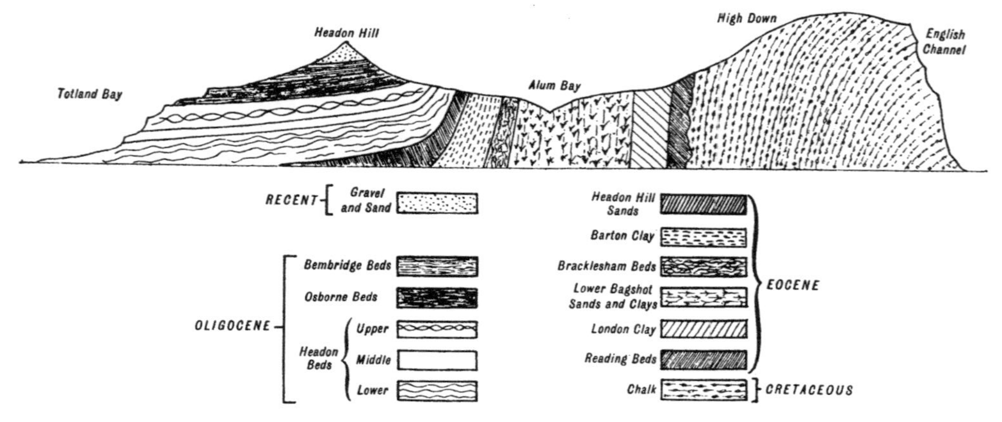

Section nearly N. to S. through Headon Hill, $\frac{1}{4}$ m. from coast at Alum Bay

Western Bight is the remarkable lustrous ochreous fluid —water highly charged with iron oxide—which exudes at intervals from the ferruginous sands and clays.

The chalk is well shown in many places—the best exposures are at Scratchell's Bay and The Needles, and at Culver Cliff. At each of these places the stratification is almost vertical and is beautifully indicated by the lines of flint.

The strata at Alum Bay are absolutely vertical. Their relationship to the other Eocene strata and to their Oligocene and Cretaceous neighbours can be at once seen from the section here given. The whole series of these rocks can be readily followed from the shore between Totland Bay and The Needles. At Whitecliff Bay is another exposure of the same series, but it is not so easy to follow nor so illustrative. Along the northern coast the Insect Limestone of the Bembridge marls in Thorness and Gurnard Bays, and the Hamstead Beds, a classic spot for fossil-hunters, must be mentioned.

The island may at once be termed a geologist's paradise and an active school of practical geology, for it affords unrivalled opportunity for the study of a number of geological processes which elsewhere have demanded long periods for their accomplishment, but which may here be witnessed in actual and at times even rapid progress. The chief of these are coast erosion, landslip phenomena, and the carving out of the chines.

The outstanding example of landslide is of course the Undercliff—the cause of which we have already indicated. But other landslides on a smaller scale occur practically every year. A great contributory cause in all these is the land-springs which break out all over the strata overlying the gault, and as these tend to change their course whenever a slide occurs it is difficult to guard effectively against them.

Not only can landslide phenomena be studied along the whole southern coast from Blackgang to Bonchurch, but their influences are to be marked wherever gault or other clays are a feature of the soil. This liability to landslides has indeed been put forward in an official application for a reduced speed limit for motor vehicles in the town of Newport, on the ground that a high speed disturbed the unstable subsoil strata and so endangered the foundations of some of its buildings.

We have referred already to the chines. With one exception, viz. Ladder Chine, these are all water-worn. Ladder Chine is an interesting example of wind action on the greensand, which has gradually worn out a kind of funnel or amphitheatre by whirling up sand, not from the beach below, but from the cliff face itself. The blown sand has steadily accumulated and now forms a line of small dunes along the cliff-top.

The water action which has worn out the chines has in general been gradual, but that at times Nature can work rapidly is shown by Cowleaze and Shepherd's Chines. Both result from a somewhat copious stream rising near Kingston village, which 120 years ago flowed along a ravine, now dry, running parallel to the shore till it found an outlet near Cowleaze Chine. A shepherd, however, one day made a small deviation in the stream, and heavy rains happening to come on, rapidly assisted the brook to cut out a new channel, and thus formed a new chine, extensive and picturesque, which is some 130 feet in height and four or five hundred feet across at the top.

The separation of the island from the mainland is another interesting problem. It appears to have occurred somewhere about the Bronze Period, partly as the result of a progressive subsidence which lowered the level of the trough of the Ancient Solent River and partly as the result of erosion of the chalk cliff which formerly extended as a continuous belt from Freshwater to Studland on the opposite corner of Dorset. This chalk belt, never very wide,

The Needles

formed a solid barrier between the sea and the Solent valley, but as a result of marine erosion along its face it grew continuously narrower, until eventually it gave way and the sea made a passage through the gap. The breach thus formed, narrow at first and quite shallow, has by the scour of the tide been steadily deepened. Probably the detached rocks, of which the Needles are the surviving examples, were much more numerous and certainly much larger.

We know from recorded observations extending over the last 150 years that the channel near Gurnard Bay has been deepened by about 10 feet. At this rate we see that it is quite possible that the separation of the island may have taken place within historic time. The present depth between Gurnard Bay and Lepe is about 66 feet, so that the old tradition of a land bridge sufficiently dry for traffic to pass over it at low water having existed in Roman times is by no means to be dismissed as impossible.

7. Natural History.

During the Glacial periods of Pleistocene days, England was united by a land connection with the continent of Europe, and no plants except those of Arctic type were able to survive these periods of intense cold on the Hampshire continent. But, with the restoration of warmer conditions, plants of the temperate zones steadily crept back, and one striking result of this return to temperate conditions, accompanied, as it was, by a decidedly moist climate, was the establishment of forests, composed of thick undergrowth and timber trees of many species, which spread right across the middle of Europe. The greater part of the South of England, right down to Roman days and even later, was covered, except here and there as to its upper levels, by a veritable jungle, and over the Hampshire continent there extended a wide woodland belt stretching east and west over the medial portion, broken only by the southern Isle of Wight area and the central chalk upland of mainland Hampshire.

The period during which the Isle of Wight has been cut off from the Hampshire mainland, though historically ancient, is not sufficiently remote to provide much variation or formation of new types, but the island has a wide range of geological strata exposed, affording much variation in country. Thus we have chalk and dry upland, sand and sandy heath, marl, loam, stiff clay, and gravel, as well as peat marshes, muddy foreshore, and blown sand, so that, in comparison with its small area, the flora is remarkably extensive.

The total number of species, apart from lichens, algae, liverworts, mosses, etc., now existing on the Hampshire "continent" (*i.e.* Hampshire plus the Island) is 1180, of which 148 found on the mainland are absent from the island. There are therefore 1032 species recorded as existing on the island, and of these 21 are absent from mainland Hampshire. The respective floras are in fact nearly identical, thus confirming the conclusion arrived at from geological and other considerations as to the comparatively recent date at which separation occurred.

Striking, however, as is this similarity, there are marked differences in relative distribution. To begin with, in the case of the timber trees, whereas in mainland Hampshire the beech, yew, and juniper are specially characteristic of the chalk uplands, on the island they are almost entirely absent. A few beech trees exist here and there, but they are quite unusual, and the all but solitary specimens of yew and juniper have probably been introduced. The Scotch fir again, which on the mainland is so prevalent over large areas, and which has within a period of about 100 years become widespread over the Bournemouth area and is spreading rapidly in the New Forest, is rarely met with in the island, except in Parkhurst Forest, where there are big plantations of it. The characteristic timber tree of the island is the elm. It is met with everywhere, and here and there individual trees attain a considerable size.

Generally speaking, the area south of the median chalk axis is bare of timber, though here and there, particularly along the gault outcrop round Shanklin, and all along the Undercliff, timber trees thrive in great variety, sycamore and ash being the chief large species. Over the gravelly and clayey soils which mark the Tertiary beds—the northern belt and the district north of Sandown—the chief tree is the oak, and oak is in fact almost universal in this district.

Before the Solent became an arm of the sea, this region, together with the areas on both sides of it, formed one continuous belt of woodland. Parkhurst Forest, the earliest recorded royal park, covered a wide belt between the chalk downs to the Solent, stretching from Calbourne to the Medina. When it was enclosed in 1815, 1150 acres were reserved by the Crown for timber for dockyard purposes. It is under the Ranger of the New Forest.

As regards the distribution of plants, the island has been divided into five principal districts, viz.:

1. The Seashore.
2. The Northern Tertiaries.
3. The Central Chalk Range.
4. The Valley of the Greensand.
5. The Undercliff with the Downs above it.

The best spot for seashore plants is St Helen's Spit or St Helen's Dover (pronounced locally, Duvver), where a remarkable profusion prevails. Here, within an area of something like 50 acres no less than 250 distinct species of flowering plants have been found growing—among them some rare and interesting species. The autumn squill (*Scilla autumnalis*), absent on the Hampshire mainland, is found in the island at this spot only, and here it grows freely, covering the turf in August and September with its blue nodding blossoms. Two very rare grasses, *Poa bulbosa* and *Festuca uniglumis*, have likewise been found here, but nowhere else in the Hampshire area. Other shore spots where special flowers are found are the corresponding sandy spit at the western end of the island, Norton Spit, the foot of Culver Cliff, Newtown Salterns, and Freshwater Cliffs.

Among the shore plants found on the island is the Samphire (*Crithmum maritimum*), formerly extensively gathered for pickling and exported. The sea cabbage, or colewort (*Brassica oleracea*), absent from mainland Hampshire, once abundant at the foot of Culver Cliff, has now disappeared from there, though it may still occur sparingly elsewhere in the island. The rare Portland spurge (*Euphorbia Portlandica*) also extremely rare on mainland Hampshire occurs at Culver, where it attains a large size; and another rare spurge, *Euphorbia Peplis*—now extinct—once reached its most easterly station in the island. More commonly occurring species are the prickly sea holly (*Eryngium maritimum*), the sea lavender (*Statice Limonium*), the yellow or horned poppy (*Glaucium luteum*), the sea heath (*Frankaenia laevis*), sea pink or common thrift (*Armeria maritima*) of which a special variety is found at St Helen's Spit, and many others. Curiously enough, some of the shore plants not uncommon on the Hampshire mainland are found but rarely on the island. Among them may be mentioned the sea kale (*Crambe maritima*), which still occurs freely at Calshot Spit (though much less so than fifty years ago), but is represented in the island only by casual specimens. This is an interesting plant, and its profusion at Calshot led Curtis, the botanist, in the eighteenth century to cultivate it and introduce it to the London market. Noticeable in the same connection is the cord grass (*Spartina*), of which three species exist. The indigenous *Spartina stricta*, though more abundant on the mainland coast opposite, occurs freely at Newtown Creek and Yarmouth, but another species, the many-spiked cord grass (*Spartina alterniflora*), probably introduced from America, established freely in the creeks round Southampton Water for about ninety years, occurs rarely, if at all, in the island. But curiously enough, a hybrid of these two kinds, the indigenous and foreign, Townsend's cord grass (*Spartina Townsendii*), first observed near Hythe in 1878, has established itself, and during the last forty years has been spreading rapidly on both sides of the Solent. The marram grass (*Ammophila* or *Psamma arenaria*) grows freely at St Helen's Spit, but is likewise less common on the island than on the mainland. These last-mentioned plants, cord grass and marram, are important as exerting a restraining influence on coast erosion and the shifting of sand dunes, and later reference will be made to them. Along the northern tertiaries, the resemblance to mainland Hampshire is naturally closer—among specially characteristic species are the narrow-leaved lungwort (*Pulmonaria angustifolia*), rare on the mainland except in the New Forest, but here abundant, a handsome plant, to be seen in full beauty in April. The common columbine, the daffodil or Lent lily (*Narcissus pseudonarcissus*), and spurge laurel (*Daphne laureola*) are common. The elecampane

Inula Helenium), very rare elsewhere, occurs plentifully at different spots in the island, notably near Quarr and Binstead. The sweet-scented butterbur coltsfoot (*Petasites fragrans*) occurs naturalised in the island, both here and towards the south. Other interesting plants found on the wet spots include the buck-bean (*Menyanthes trifoliata*) and the round-leaved sundew (*Drosera rotundifolia*), though neither is common.

On the central chalk range we find of course different species. Near Ape's Down, almost its only British station, the wood calamint (*Calaminta sylvatica*) grows freely. The toothwort (*Lathraea squamaria*) occurs growing parasitic on hazel, etc., and the black mullein (*Verbascum nigrum*). Here, too, is a whole series of the common characteristic plants, such as rock rose (*Helianthemum vulgare*), salad burnet (*Poterium sanguisorba*), etc., and, in addition, many kinds of orchids.

In the Greensand Valley we find a large range of heath plants, in the wet valley bottoms a number of characteristic bog species—several heaths, the common foxglove, starworts, pasture lousewort (*Pedicularis sylvatica*) and many others.

The Undercliff and downs above it again afford a specially wide and attractive flora, and all along the Undercliff adventitious or introduced plants find conditions highly favourable for their establishment. The mountain St John's wort (*Hypericum montanum*), purple cow wheat (*Melampyrum arvense*), the very rare milk vetch (*Astragalus glycyphyllus*) and the stinking hellebore (*Helleborus foetidus*) are found here. The very rare spider orchis (*Ophrys aranifera*) was once found here, but no recent record exists of its occurrence. The red valerian, an introduced plant, grows freely at many spots, and such garden plants as hydrangea, fuchsia, etc., are common. The bilberry (*Vaccinium Myrtillus*) grows along with heather, dwarf furze, and blackberry all over Boniface and Shanklin Down. A variety of broomrape (*Orobanche Hederae*), parasitic on ivy, occurs here exclusively. The Italian cuckoo-pint (*Arum Italicum*) occurs at Steephill and practically nowhere else. At Niton an interesting plant, the sweet galingale or Cyprus grass (*Cyperus longus*), grows commonly. The most striking external characteristic of the Undercliff, however, is the variety and beauty of the woodland which stretches almost continuously along it, and after a break continues on round the hillside behind Shanklin. An outstanding feature of the island generally is the abundance of the blackberry.

Orchids are well represented, particularly on the dry chalk uplands. The lady's tresses (*Spiranthes autumnalis*), with its little spikes of minute fragrant blossoms, is usually to be found freely over the open pastures, as for instance, by Chale. As a result of the scarcity of beech trees, the large white helleborine, so characteristic of the beech woods of mainland Hampshire, is unknown in the island. The curious "bird's-nest" (*Monotropa Hypopithys*), which, like *Neotti*, the bird's-nest orchis, haunts the beech woods of mainland Hampshire, is also but rarely met with.

Of ferns there are several interesting kinds, though the actual number of species is not great, tourists and others have uprooted and in fact entirely destroyed them in many places. The ceterach occurs at Brading and Carisbrooke, and a few other places, the maidenhair (*Asplenium trichomanes*) in some plenty at Quarr and elsewhere, and *Asplenium rutamuraria* and *A. adiantum nigrum* are also found. The royal fern (*Osmunda regalis*) has an especial attraction for plant robbers, but, though nowhere common, is still to be found at a number of places.

The bird and animal life of the island is rich and varied, though, with the increase of its human population, many interesting birds have disappeared. The high cliffs of Culver, the Undercliff, and Freshwater, are still a refuge for birds, such as the cormorant, the shag, razorbill, guillemot, puffin, herring gull, and many other types of sea-fowl. Jackdaws innumerable nest all along the Undercliff, and up to some 60 years or so ago the chough was still to be seen. The raven and peregrine both still nest on the Freshwater cliffs. The stock of ravens was renewed with fresh blood from Yorkshire, somewhere about the year 1909, when there was only one native bird left.

Among residents, in addition to the above, may be noted the kestrel, sparrow hawk, several species of owl, jays, magpies and goldfinches. The heron, common on the mainland, rarely nests here, though it is a frequent visitor on the shores and mud flats. The kingfisher, once common, is now also rare.

Summer migrants are numerous. The wheatear and ring ousel are among the earliest arrivals. The nightingale is specially abundant. The corncrake was formerly much more numerous than it is now, but it still breeds here occasionally. Among winter migrants are the golden plover and many sea-fowl, as the Brent (or in the island Bran) goose, the great northern and black-throated divers, etc.

Very numerous indeed are the occasional visitors. The last white-tailed sea-eagle is dated as far back as 1780, but the golden eagle has been recorded quite recently, though possibly specimens seen may have been escapes from captivity. The hobby, osprey, hen harrier, buzzard, marsh harrier, and honey buzzard are among these. A beautiful specimen of the white stork was captured at Shorwell in 1902. The hoopoe is a casual visitor, as is also the quail.

Modern encroachments and modern conditions, however, tend rapidly to restrict the number of these more picturesque bird forms, and while human residents and human summer migrants do their work of extermination on land, the oil accompanying the presence of motor craft on the inland waters is an equally destructive agent both for the birds and fish which haunt them. Great numbers of sea birds killed by this agency have been recorded, especially within the last couple of years.

Of land mammals the range is considerable. Bats are specially plentiful. They include the greater horse-shoe bat, the long-eared bat, the serotine and others, as well as the common bat. Hedgehogs are abundant. The otter is occasionally seen. The badger, once not uncommon, may now be regarded as extinct, and the red deer, once hunted in Parkhurst Forest, disappeared long ago. The fox, now plentiful, appears to have been quite unknown 120 years ago.

The island waters have also their interesting denizens. Among aquatic mammalia, the porpoise (*Phocaena communis*) is not infrequent, and makes its way at times up the Medina to Newport itself. Dolphins are occasionally seen. The common seal is recorded as having been last seen about eighty years ago. Whales of various kinds are occasional visitors. A specimen of Rudolphi's rorqual, 39 feet long, was hunted and driven ashore at Seaview in 1888. A common rorqual (*Balaenoptera musculus*) was stranded in

1842 at Totland Bay, and its skeleton, some 80 feet long, is one of the curiosities exhibited at the so-called "museum" at Blackgang.

The land reptiles include the common (viviparous) lizard, the slowworm, the grass snake, and the viper. The last named is very common still, though its numbers are decreasing; the grass snake, however, is comparatively rare.

8. The Coast. Tides. Erosion.

The coast of the island presents much variety—bold cliffs, broad sweeping bays, low flat clay shores, winding creeks, and desolate mud flats are to be found. Both in form and colour the range and variety of type are striking.

The loftiest part of the actual coastline is at Main Bench, Freshwater, which rises nearly 500 feet, and this is rivalled by the whole escarpment of the Undercliff—in essence the true coast, although the slipped material which terraces its face guards it from actual contact with the sea. This escarpment is perfectly vertical, and is fringed at the top with deeply weathered and often projecting bands of chert and freestone, the specially hard nature of the former enabling it to resist the wind and weather, which pick out the softer material between. The actual top of this escarpment is formed by chalk which caps the greensand over most of its upper levels. This chalk forms downs, rising at St Catherine's Hill to 781 feet, at Week Down to 692 feet, and at St Boniface Down to 787 feet. The actual vertical face of the escarpment is nowhere so high as this,

Gore Cliff, near St Catherine's Point
(Showing weathering of ragstone at top of cliff)

but reaches 540 feet at Gore Cliff, and 474 feet at High Hat near St Lawrence.

Although all along the face of the Undercliff the sea washes the base of false cliffs of slipped chalk—their strata so undisturbed as to appear like true cliffs *in situ*—the chalk formation actually reaches the sea only at the ends of the

Scratchell's Bay

great median axis, viz. at Culver Cliff to the east, and Scratchell's Bay to the west. At both these points the cliffs are very fine. At the western end the three detached chalk masses—the Needles—and the Arched and Stag Rocks in Freshwater Bay tell their tale of coastal wear and loss. Southward between these two bold extremities, the coast forms a series of curving bays, the Western Bight, as for want of a better name we have termed it, on the west, and the beautiful and well-defined Sandown Bay on the east. Between these are a series of coves at the base of the slipped chalk of the south, Puckaster Cove, Steephill, and Ventnor Beach, and Luccombe Bay beyond.

The cliffs along the Western Bight maintain a fairly general level of from 100 to 200 feet, only falling to about 50 feet at Compton Bay. Here the chalk ceases and the remainder is largely a continuous vertical wall of sand, with masses of talus and slipped material at intervals along the base. The beach is mainly fine shingle, with a narrow belt of firm sand at low water. The chines along its course have already been spoken of. Several give direct access to the beach, but others such as Walpen and Blackgang Chines terminate at a considerable height above it. Much alteration has occurred in the form of Blackgang Chine during the last hundred years or so, and a comparison of its present appearance with that as depicted in old prints is instructive.

This part of the island coast is particularly dangerous to shipping, partly from its liability to sea fogs, and the variable set of the currents, and perhaps most of all to a broad submerged shelf or ledge fringing the shore in front of Atherfield—revealing itself at low tide by the colour of the sea and a slight "lip" or curl of the water as the waves wash over it. Chale Bay is indeed regarded as one of the most dangerous spots on the whole British coast, so much so that as many as fourteen vessels have been known to be wrecked here in a single night. One such wreck, as far back as A.D. 1312, had as a curious consequence the erection on the top of St Catherine's Hill hard by, of a pharos or lighthouse, and of an oratory or hermitage adjoining it, with a priest in charge whose principal duty was to attend to the

Walpen and Ladder Chines, near Chale

beacon light—an act of enforced piety on the part of the local Lord of the Manor who had illegally possessed himself of the cargo of the wrecked vessel, and who was condemned by ecclesiastical authority to make restitution in this manner. The tower still stands, but the oratory has disappeared, and the 14th-century pharos is now represented by the magnificent St Catherine's Lighthouse, erected in 1840.

Culver Cliff, Sandown Bay

Chale Bay is exposed to the full force of weather and the whole sweep westward of the Channel; it witnesses seas far grander than at any other part of the island coast. The waves as they break and recede from the shingle exert at this point a very marked suction effect or undertow, so that bathing here is more or less dangerous. The foreshore of the Bight at the Freshwater end is strewn with big and partially worn flints. It is curious to notice how these are continuously rolled eastward by the drift of the tide, and how rapidly they grow rounder and smaller in the process. Similarly the red shingle formed by cliff waste grows steadily lighter in colour as well as finer in bulk in the same eastward drift. This bleaching is caused by alternate oxidation and solution, the result of repeated exposure to air and to sea-water.

East of the island, Sandown Bay, in some ways equally

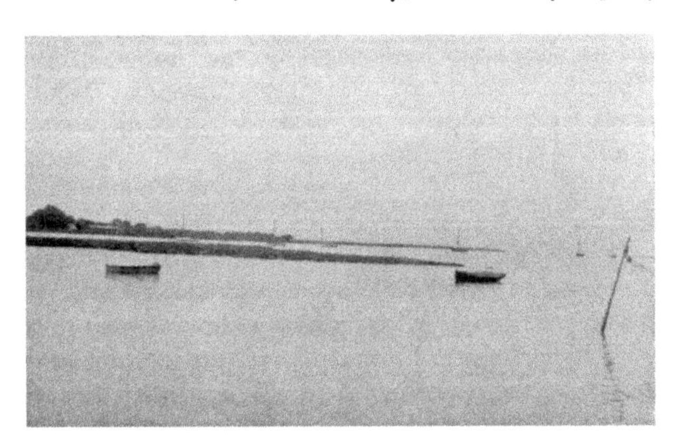

Newtown Estuary looking seaward
(Oyster beds along Shallows)

picturesque, offers fewer points of interest. It presents a grand sweep of cliff and beach from Dunnose to Culver though rather spoiled by the quantity of rough boulders at the Shanklin end, which recently have tended to increase both in size and number. The cliffs above Shanklin—Dunnose and Luccombe in particular—are grand and forbidding. Luccombe and Shanklin possess chines each bearing their respective names. Along Sandown Flat between

Sandown and Culver the land descends actually to sea level, and this part of the coast has been artificially protected by groynes and sea walls.

Of quite different character is the northern coast from, say, Foreland to Alum Bay. For the most part it is composed of stiff clays, varied here and there by some limestone, almost everywhere of low elevation, and frequently verdure clad almost to the water's edge.

As already noted it is characterised by deep, branched, and winding creeks—Wootton Creek, Medina Estuary, Newtown Estuary, and the Western Yar—in which list Brading Harbour, now reclaimed, had formerly to be included. Of these the Medina Estuary is the only one now carrying any bulk of commerce, though at Wootton Creek and Yarbridge some seaward trade is carried on.

It was Yarmouth and the Western Yar which inspired Tennyson's *Crossing the Bar*, and seen indeed as the poet probably saw them when crossing, as he was wont to do, from Lymington to Yarmouth, with the sun low on the horizon or the deepening incertitude of twilight, they suggest an aspect and quality of island scenery rarely perceived by the casual visitor. To those who know and love the island, Yarbridge and Thorness Bay have as much a place in the picture as Chale and Blackgang, Whitecliff Bay and Culver.

To appreciate the special tidal phenomena presented by the Isle of Wight waters, and nowhere else in the British Islands, we must study the tidal action of the English Channel as a whole.

Local conditions are very complex: one main principle operates everywhere—the law of Eastward Drift. The tidal wave rising in the Atlantic causes a broad swell over

Yar Bridge and Causeway

the wide mouth of the Channel from Brest to the Land's End, which passes up Channel as up a narrowing funnel, each point of the coast serving, as the wave reaches it, as the centre of a fresh tidal pulse, in accordance with the well-known laws of wave motion. Eastward the wave sweeps with a scour that grows in intensity the further it advances, carrying with it the loose stones, shingle, and sand that lie round the shore, and cutting into the base of the foreshore as it goes. When the tide falls the conditions are reversed, but the receding waters flow from a narrow channel towards an ever-widening one, and though the scour westward serves to some extent to move the sand and shingle back again, it does so with diminished force. Thus, every tide tending to carry eastward more material than it brings back, the detritus is continually being transported eastward.

The material of the beach at any particular spot will depend on the kind of rocks to the westward of it, but its coarseness or fineness depends on the intensity of scour. The coarser the material the more rapidly it is deposited; the finer it is, the longer it remains in a state of suspension, only settling in comparatively smooth water. At a point or headland directly in the line of scour we shall perhaps find no beach at all; where the scour is less direct we shall find shingle; and only in sheltered areas where the flow is very slack shall we find fine sand or mud. Thus the parts where there is rapid scour are points of erosion, and those of quiescence tend to be points of deposition, and all along the Channel the estuaries and lagoons tend to silt up and sand-banks to form in quieter spots away from shore.

In the inland waters of Hampshire we find a different set of tidal conditions from those prevailing in the open Channel. We have what is commonly termed the "double tide."

The general tidal movement making from the westward is split at the Needles, and while part of the flood passes up the Solent, the main body sweeps round the south of the island and reaches Spithead from the east, reacting with the other body of water and thus causing two periods of maximum high water. The intervals between the two maxima are as follows:

Solent.	*Spithead.*
Cowes, 60′	Hythe, 77′
Beaulieu, 110′	Southampton, 97′
Lymington, 110′	Redbridge, 110′
Yarmouth, 120′	Calshot, 55′
Hurst, 120′	
Christchurch, 150′	

During the period between the two maxima the variation of depth is very slight—thus at Portsmouth the time required for a variation of one foot at high water is 2 hours, and at Lee 2½ hours, but at Calshot, Hamble, Netley, Hythe, and Southampton it is about 3½ hours. These figures show us that while in Spithead the effect of this double tide, as it is called, is merely that of prolonged high water, along Southampton Water and the Solent two definite maxima are reached, the intervals between them increasing as we get farther from Spithead. Nor is the effect of this confined to the Solent; it is felt, though with diminishing intensity, as far along the coast as Portland, the interval increasing progressively the farther west we go. At Weymouth the double tide corresponds with low water instead of high water, and Weymouth has therefore a double low tide, known locally as the "Gulder."

The double tide, by prolonging high water in Southampton Water and the adjacent channels, has been a material feature in the commercial development of the port of Southampton. It has been noted as a peculiarity of our coast by the earliest observers, and Bede, writing in the eighth century, gives a description of the phenomenon which is as correct as it is vivid.

While little can as a rule be done permanently to arrest erosion, it can often be delayed and sometimes entirely checked in certain spots by building sea walls and groynes. The latter are barriers of timber or stone running out to seaward from the foreshore at right angles. The sea-borne detritus, *e.g.* shingle and sand, collects on these groynes, heaping itself at first up the western side (law of Eastward Drift) until perhaps the whole is finally covered. When this is the case the groynes remain as a permanent barrier guarding the foreshore from erosion at its base. It will thus be seen that the cartage of shingle from the foreshore is highly inadvisable, as it removes its most efficient protection.

Erosion, actual or potential, is active all along the island coast, though its intensity varies greatly from point to point. It is most active along the northern shore, though at the north-eastern corner, between East Cowes and Ryde, it is much less than elsewhere. This erosive activity is caused largely by the rapid scour of the tide, particularly between Sconce Point and Hurst, where the main flow passes along at six knots an hour. At Yarmouth the town itself is protected by groynes and sea walls, but the scour has cut back no less than 150 feet along the common in some 50 years, and this portion is now protected by a low sea wall with a curtain or "apron" of shingle in front, while west of Gurnard Bay the cliffs are being cut back about 5 feet a year. Erosion along the northern shore is to some considerable degree accelerated by the wash of large steamers proceeding along the Solent. Not only does the tidal scour tend to widen "the River," as the sailors term it, but every year it gets deeper. Such data as exist seem to show that this is going on at a rate of about 1 foot in 15 years (see *ante*, p. 34).

The intensity of erosion has in the past been accentuated by the imprudent removal of shingle from the foreshore—especially at Culver Cliff and at Freshwater, the two real danger spots of the island. At Sandown Flat the danger appears now to have been entirely obviated by an elaborate system of groynes and sea walls. At Freshwater, however, a different set of conditions exists. Here, as at the Culver end, a stretch of land practically at sea level, Freshwater Gate, divides Freshwater Down from the main part of the island, and the danger of the sea breaking through is here accentuated by the special tidal circumstances. Under ordinary conditions the rise of the tide along the Solent is liable to cause a dangerous swell up the Western Yar, and if this is aided by a strong following wind or unusual tidal height—a circumstance recurring practically every winter—there is danger of a break occurring through the Gate, all the more so as the alluvial material which forms its base would be incapable of offering much effective resistance to being swept away. But another danger due to the "double tides" is also present. The full tide rising in the Channel divides at the Needles, and sweeps along the island on both sides, by Yarmouth to the north, and Freshwater to the south. Two hours afterwards (see table on p. 53) the tide is again high at Yarmouth, while at Freshwater it has fallen, and thus, quite independently of any reinforcement by wind, there will be normally several feet difference of

The Groynes at Sandown

level between the water heaped up in the Yar Estuary and that on the coast at Freshwater. The whole matter has formed the subject of the Royal Commission on Erosion.

St Helen's Tower, Brading

Though groynes and retaining walls have their value, especially when screened by shingle, they are, of course, useless against shore drainage and progressive sliding over sloping moist clay surfaces. An illustration of this is the cracked masonry and the uneven and broken slabs of the esplanade and sea wall at Egypt Point. More striking evidence of what erosion can do is possibly the half-ruined tower of old St Helen's Church, left as a beacon when coast erosion had necessitated the pulling down of the rest of the structure.

Wherever the foreshore is muddy or lined with sand dunes much protection can be afforded by promoting the growth of such shore plants as the Spartina and marram grasses mentioned in the preceding chapter.

The one outstanding instance of island reclamation—the converse of erosion—is that of Brading Harbour, already referred to. The Erosion Commissioners reported that further reclamation at any point of the island was not practicable.

The navigation of our Hampshire channels and waterways is greatly affected by the sandbanks and their tendency to shift, and were it not for the special tidal conditions already discussed Southampton Water would very possibly be comparatively unnavigable. As it is, its deep water channel or fairway (5 fathoms) contracts in places to very narrow dimensions. Off Calshot itself it is barely a quarter of a mile wide, and from that point it takes a course south-west, so that ocean-going steamers leaving Southampton Water bear right over towards Cowes whether proceeding down Spithead or the Solent. All through Spithead sandbanks occur, marked by bell buoys or lightships, and off Portsmouth there are others, on some of which forts have been erected—Horse Sand, No Man's Land, Spit Fort, etc.

Lighthouses and lightships are numerous; they include the Needles, Egypt Point at Cowes, Hurst, Calshot, St Catherine's, and the well-known Warner, and the Nab. Both the Needles and St Catherine's are occulting.

9. Climate.

While by "weather" we mean the particular atmospheric conditions of the moment, the word "climate" sums up their general tendency and results, together with the seasonable changes usually experienced over a series of years. Temperature, wind, rain and snow, dew, cloud, humidity and sunshine are among the chief factors in connection with climate.

In England by far the largest amount of precipitation is in the form of rain, though snow and dew are important also, and particularly so in agriculture. Precipitation is greatly promoted by cooling of the atmosphere, and one important cause of this is the well-known physical law that if air passes into a space where it can expand temperature immediately falls and it parts with its moisture. Elevated land masses by deflecting air currents upward thus exercise an important influence in promoting precipitation. Moreover, movements of air circulation usually take the form of rotary systems. These are of two general kinds, viz. cyclones, in which the region of lowest pressure is in the centre, and anticyclones, in which the highest pressure is in the centre. The former are associated with rainfall, the latter with dry weather, so that the passage of a cyclone or an anticyclone over any district brings rain or drought quite independently of any influence of land masses. With us cyclones are associated with south-west winds, and as the chief elevations in the British Isles are in the west, it will be seen that as we pass from west to east the rainfall becomes

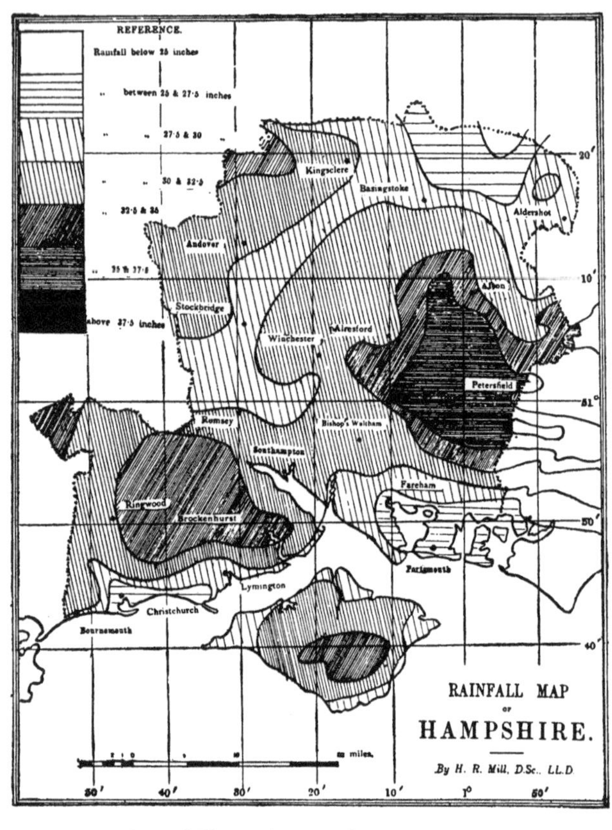

Rainfall of Hampshire and the Isle of Wight

(By permission of Dr Mill and of the Controller of His Majesty's Stationery Office)

less and less, varying from over 80 inches a year in Killarney and Glamorgan (it actually reaches 120 inches at Borrowdale in Cumberland) to 40 inches over Salisbury Plain, and from 25 to 37½ inches over Hampshire and the Isle of Wight, down to under 20 inches in Cambridgeshire and parts of the East Coast.

Over a small region like the Isle of Wight no striking variation of rainfall is to be looked for. It averages from 27½ to 35 inches per annum, and curiously enough the rainfall is greatest in the centre, *i.e.* over the "Bowl," and diminishes as we approach the coast. The general distribution can be clearly seen by reference to the map. The prevalent wind, the south-west, is also the humid one, and as on reaching the island this is sharply deflected upwards by the southern heights precipitation is induced, but the sweep of the wind carries it forward before the full effect is secured, and so it is the "Bowl" beyond, and not the heights themselves, on which the greatest amount of rain descends. As a further illustration of this, even where the weather is otherwise quite fine, one or other of the island heights, such as St Catherine's Down, will be seen covered with a mist or cloud-cap—"putting on its nightcap," as the island folk say. This is the first stage of precipitation, and if the process is not carried beyond this stage the mist is merely persistent over the hill and no rain actually falls, but if rain does fall it may only begin to do so a little farther away, where the sweep of the wind has since carried it. This tendency to mist is a general feature of the island climate, and climatically its chief drawback. Chale mists are proverbial.

Following the general laws of climate, however, unusual visibility is noted in the island as a presage of rain—thus when Portland is visible from Chale one is told to look out for rain, and another local indication connected with the same spot is linked up with St Catherine's Tower itself. The upper part of the octagonal tower is pierced by eight openings, and when anyone at Chale Green can see daylight appearing through these openings, it is looked upon as a sign of approaching rain. As a result of this tendency to mists, halos and bows are of frequent occurrence. Solar halos have been recorded as being observable annually on 15 days, and lunar halos on 3 days. Fogs are stated to occur on an average on 29 days in a year. The general humidity is decidedly greater too than over the adjacent mainland. In spite of this the island is one of the sunniest spots in the whole of Britain, and this is particularly the case in the first quarter of the year.

In an average year the island rainfall reaches its minimum in the spring, and its maximum six months later. Observations over 22 years, at St Lawrence and at Osborne, both give May and October as the months of minimum and maximum fall, the ranges being:—

St Lawrence: May 1·57 inches, October 3·66 inches.
Osborne: " 1·61 " " 3·39 "

As regards temperature there is much local variation—aspect, elevation, soil and subsoil, and particularly the amount of moisture in the subsoil, are all important conditioning factors. Owing to the dry and sandy soil conditions, Sandown and Shanklin enjoy a specially warm and genial climate, while the sub-tropical conditions prevalent along the Undercliff are in the main attributable to aspect. In the north, over the moist soil of the Oligocene and the Tertiaries, as at Cowes and Yarmouth, more humid conditions, marked by greater cold and numerous sea fogs, prevail.

But mild as is the island climate in general, severe winters have been from time to time recorded. That of 1798–99 was phenomenally so—the island lay under thick snow for weeks, cattle and sheep in large numbers perished, and many people even were frozen to death.

10. People. Population. Dialect.

Of the earliest types of man who inhabited our land —men of the Palaeolithic and Neolithic ages—the traces left on the island are but scanty. The most ancient of all these is known as the Piltdown man—*Eoanthropus Dawsoni*, or the man of the dawn—whose skull was found in 1912 at Piltdown in Sussex, and from the proximity of the island area to Sussex it was probably men of his type who first trod its soil. Palaeolithic man was a hunter and wanderer, and he made his way to Britain over the land bridge which in his day joined Britain to the Continent. He had no knowledge of agriculture and no domesticated animals—his tools and weapons were of flint rudely chipped and fashioned. The island of course then formed part of the mainland, but his presence here was probably little more than casual, and the dense forest which covered the whole of the north, and the swampy jungle of the Bowl would offer little facility for a hunter to find subsistence.

Neolithic man, who after a long period replaced him, led a more settled life. He dwelt in general on the upper levels, where he pastured his flocks and herds and probably carried on primitive cultivation. In person he was short, and had a dolichocephalic (*i.e.* long and narrow) head, and he carried the art of fashioning stone implements to high perfection.

The chief spots where he lived in the island seem to have been near the shore—a number of shell heaps of the kitchen-midden type have been found in the Undercliff and elsewhere which indicate his presence—but in any case his occupation was not important. The Longstone on Mottistone Down—two great stones, one erect and one

The Longstone, near Mottistone

horizontal—may possibly be the remains of one of his dolmens, and if so would be of the highest interest as being the only known Megalith occurring locally, but the stones are not now in their original positions and evidence as to their origin is too scanty to be of any service. We do not know how long the Neolithic period lasted, but it was certainly after the land connection of Britain with the Continent had been severed.

Then, somewhere possibly about 4000 years ago, a different race came on the scene—a race stronger and more developed, with a skull rounder and broader (brachycephalic), a man who knew the use of metals. We call him the Bronze Age man. It was apparently about this period that the island came to be separated from the mainland.

Later came men of Celtic origin, who knew the use of iron, followed by Romans, Anglo-Saxons, Danes, Normans. At the time of the Romans, Hampshire and the island were occupied by various Celtic tribes—Belgae in the south and in the island; Atrebates and Segontiaci in the north.

Then came the Romans, and with them the first definite mention of the island in history. The Emperor Claudius in A.D. 43 sent an expedition under Aulus Plautius to conquer Britain. Suetonius tells us "He fought 30 battles with the enemy. He reduced to submission two highly powerful nations, 20 towns, and the Isle of Wight, contiguous to the mainland."

But though there are abundant traces of the presence here of the Romans, the island possessed very little importance in their eyes, having no strategic value.

The racial impress which both they and their predecessors have left on the island has been nil. It was from Teutonic races which from about 450 A.D. onwards invaded the land that the present island stock derives its origin—probably first Jutes and later Gewissae or West Saxons. Sailing up the Channel, originally for mere raiding or piracy, but later for settlement, these folk, after the Roman legionaries had departed, speedily made the island unsafe for peaceful occupation. The scanty island population of Britons and Romano-Britons melted away, their place was taken by Teutonic settlers, and these formed the parent island stock. As to how far the Jutish element persisted as an influence in moulding the future population is a problem which cannot be discussed here. For though they settled in the Isle of Wight, and at one or two spots on the Hampshire mainland, the Gewissae ultimately established themselves as the predominant Hampshire race. The references of Bede and of the English Chronicle to the subsequent struggle between these two peoples are hard to follow and to fit in with the archaeological and other evidence. The Chronicle tells us that "the West Saxons under Cerdic in A.D. 530 overran the island with great slaughter of its existing occupiers at Wihtgarasbrig (Carisbrooke) and that Cerdic handed the island over to Wihtgar." We have no further mention of the island till A.D. 681, at which date Bede tells us that "Wulfhere of Mercia gave to Aethelwalch of Sussex, on the occasion of the baptism of the latter, two provinces, viz. the Isle of Wight and the province of the Meonwara," and further that "in A.D. 685 Caedwalla, King of Wessex, slew the same Aethelwalch and conquered Wight which hitherto had been wholly given to idolatry." While Bede's account of these later happenings is doubtless reliable, it is difficult to piece in with it the reference to Cerdic's conquest of the island and the slaughter of Jutes in 530. What is clear is that a bitter struggle for the island occurred between the Jutes and the West Saxons and that the latter ultimately became predominant.

From this time on to the Norman Conquest life on the island was insecure owing to piratical attacks by the Danes, but the latter never settled here, and the next actual settlers were the Normans at the time of the Conquest, when they dispossessed the Saxon landowners, and a great infusion of Norman characteristics resulted. From that time there has been no change of island stock except as the result of interaction between island and mainland, and recent census results seem to show that this interaction has not been sufficient in the past to check the results of segregation and inbreeding. Modern conditions seem likely

Town Hall, Ryde

however speedily to obliterate such small definite distinctions of race and type as do still continue.

The total population of the island in 1921 was 94,697, an increase since 1911 of 7 per cent., viz. 6115 persons. The density of population is 642 per square mile, as against an average density over the mainland of England of 649 per sq. m., and a maximum density (in Lancashire) of 2656 per sq. m.

But while the general density of population is thus seen

Town Hall, Newport (*Built* 1814–16)
(*From an old engraving*)

to be normal, its distribution is very uneven. Municipal boroughs and urban districts covering only about one-thirteenth of its area include no less than two-thirds of the population, and the average densities over urban as against rural areas is as 25:1.

There are two municipal boroughs—Ryde with 11,295 and Newport with 11,036. The urban districts are Cowes 9998, Sandown 7664, Shanklin 7374, Ventnor 6063, St Helen's 5706, and East Cowes 4636. Growth at Shanklin is very rapid, and the population has increased in 10 years at the enormous rate of 55 per cent. Sandown rivals this with a rate of 38 per cent. Newport and East Cowes are practically stationary, one with an actual decline of 118 and the other a decline of 23. The rural population is almost stationary.

The language of the island is closely allied to the dialects of adjacent Hampshire and Dorset; which are indeed the survivals of the standard dialect of the Old English race. The vowel system of the West Saxon tongue has been shown recently to have been practically identical with that of the earlier West Germanic parent tongue, and a number of its so-called provincialisms reflect in reality characteristics of the early English literary speech—the language of Alfred and the scholars of his day. It is interesting to notice that the island speech differs markedly from that of Sussex on the one hand and that of Devon on the other, while contrary to what might have been expected there is scarcely any trace of French influence discoverable in it.

But while, in general, dialect peculiarities nowadays tend rapidly to become obliterated, definite characteristics of race and language are still to be met with. The islander, like other island folk, still tends to be clannish, and practically lifelong residence in the island is not considered sufficient by island-born folk to remove from the immigrant from the mainland the taint of being an "overner," *i.e.* one from over the water.

11. Agriculture.

Agriculture is the most widespread industry in the island in the sense that a larger number of persons are actually employed in agriculture than in any other specific industry, but inasmuch as the urban population tends to increase and the rural to decline, its preponderance relatively to other industries tends to diminish. As the proportion of rural areas to urban is 12:1, it will be seen how predominant a position agriculture occupies in the Isle of Wight.

The general distribution of agricultural activities can perhaps best be appreciated by comparison with mainland Hampshire.

While the general proportional acreages under cultivation are the same, there is in the island relatively only one-third as much bare fallow. As regards crops, the figure for wheat is the same over both areas, for barley twice as high in the island, for oats a little less, and for roots decidedly higher. As regards live stock the comparison is greatly in favour of the island. Horses, cattle, and sheep are all relatively about twice as numerous and pigs nearly four times. Accordingly while arable cultivation is well developed, pasture and dairying are the main features of island agriculture.

As regards soil there is extreme diversity, and prolonged personal experience of a particular locality is considered in the island to be a *sine quâ non* before one can be in a position to decide as to the most suitable type of farming for any particular farm. There is of course a marked difference of soil between the two natural areas separated by the median chalk ridge. Generally speaking, over the north, dairy-farming largely predominates, and arable over the southern, but there is much local variation.

Dorset Horn Shearling Ram

Although in the past the island has had a great reputation both for wool and for wheat, agriculture over the north was very backward up to the middle of the 19th century. The heavy, cold, wet yellow clays or clayey sands and gravels were at that time largely undrained and therefore unworkable, and the methods applied were hopelessly out of date. In his classic Report on Agriculture in Hampshire and the I. of Wight, Wilkinson, in 1861, says "It is not too much to say that the north of the island is as a whole a century behind. One would suppose things could not go on as they are if they had not gone on so long already. With an open fallow, with 14 bushels of wheat per acre, with no roots and little stock, the present system would seem, in these days of competition, doomed, in spite of the milkpail." It was the Prince Consort's model farm at Barton, where draining and other up-to-date methods were

Dorset Horn Sheep, Atherfield

efficiently carried out, which pointed the way to improvement. Now the island abounds in signs of skilful and successful agriculture, as anyone visiting Newport market on a Tuesday can easily appreciate.

On the dairy farms, as a rule, about two-thirds of the area is used for milk and butter production and a limited area for rearing heifers. Grazing is done only on a small scale, *e.g.* as at Sandown Flat. The pasture is good and cattle can be fattened in many parts without the use of cake. The chief breeds are crossbreds, chiefly shorthorns and Guernseys, though in some dairies here and there good pure shorthorns and pure Guernseys are kept. The Isle of Wight milk has a high reputation and considerable quantities of island milk are sent to Portsmouth.

As regards sheep the chief breed is the Dorset Horn, and the island is noted for many registered flocks. Indeed the island flocks rank among the best of this breed, a special

Tudor Stone Cottages, Chale

feature of which is their early lambing. The lambs are dropped early in November and fatten quickly, so that grass-fed lamb is available for the London market as early as March or April. It is on the west and south-west of the island that sheep grazing is mainly followed.

Dorsetshire, as the name implies, is the home of the breed, but island-bred rams are in great demand for stock purposes both for home and for export, and a good number are sent every year to the big sheep-fair held at Dorchester.

Island pigs are a special feature. There are two leading breeds, the so-called Saddle-back and the Large Black.

The former, readily distinguished by a black body with a saddle-shaped marking of white over the shoulders, is an unusually interesting variety, which has received special attention during the last few years from the Wessex Pig Society. It is claimed for this pig that it is an old breed indigenous to the New Forest, but in spite of controversy on this point, it is undoubtedly one of the most useful breeds. Its length and growth make it a valuable bacon pig, as it is said that it can be brought on ready for market earlier say by about five or six score than other breeds. There are several registered herds of these pigs in the island. The Large Black also has its supporters and is popular as a useful and profitable breed.

For horses the island again has a decided reputation. The chief type bred, the Shire Horse, can hold its own against horses from any other English county except perhaps those of the fens of Cambridgeshire and Lincolnshire. A committee of farmers, the Isle of Wight Shire Syndicate, has of late years done much to improve the quality of the breed.

The best corn land is found in the southern part on the upper levels. Oats occupy the largest acreage, but barley is the crop for which the island is specially noted. The sea fogs are highly beneficial to it, and barley of the highest class, as judged by quality and by colour, can be grown throughout the whole island. Roots are important, the mangolds as winter feed for dairy cows and the swedes and turnips for feeding off by sheep.

Few special crops are grown, *e.g.* a little maize, and a small quantity of linseed. Potatoes occupy only a small acreage, and orchards and small fruit are only of small extent.

Poultry, especially ducks and geese, are abundant everywhere. As regards bees, the term "Isle of Wight" is one of foreboding on account of the widespread havoc caused ten or twelve years ago by the dreaded bee malady—*Nosema apis*—popularly spoken of as "Isle of Wight disease." Only a few stocks as yet have been re-established in the island.

Timber of course is mainly confined to the cold heavy soil of the north. Parkhurst forest is still the chief expanse

Chalk Quarry, Brighstone

—it is chiefly oak and fir, and none of it is of large size. The fishermen make their lobster-pots from osiers grown here and there in the swamps.

A prominent feature of the chalk downs, especially at the higher levels, are the large chalk quarries. "Marling" is of very ancient origin and is still practised. The best chalk for marling comes from the median ridge, as it pulverizes readily under frost. The southern chalk mostly flakes up but does not so easily pulverize, though that of Ventnor readily crumbles.

There is only one stock-market in the island, viz. Newport market. Every Tuesday the roads leading to this town are thronged with vehicles, cattle, and every kind of market traffic. For many island folk Newport market is the one serious event of the week and Newport itself on market day is a scene of much bustle and picturesque activity.

12. Industries and Occupations.

After agriculture, the best defined occupations of the island are engineering, and land and marine transport. A comparatively large number of persons are engaged in official occupations (Parkhurst Convict Prison accounts in part for this) and in naval or military service; according to the last census, 741 on the former and 2400 on the latter. Other classified occupations for male workers over 10 (total number 33,600) are:

Agriculture	3748	or	15·1	per cent.
Railway, land and marine transport ...	3545	„	14·3	„
Food, tobacco, drink, and lodging industries	2900	„	11·9	„
Engineering, including men on ships and boats	2822	„	11·4	„
General (including factory) labourers ...	1529	„	6·1	„

These figures, while they show the leading part played by agriculture and the importance of transport (land and marine), show also how large a place skilled engineering occupies in the dominating island occupations.

Of the three leading urban centres, Ryde, Cowes, and Newport, Ryde is the leading one for transport and packet services, and Cowes for marine engineering, yacht-building, and yacht equipment; while Newport, in addition to acting as the seat of official government, the stock and produce market, and centre of distribution, carries on a considerable timber, brewing, and milling industry.

It is on the Medina area between Newport and Cowes that industrial activity is centred. At Dodnor are the West Medina Cement Works, the tall chimney of which is visible for miles round, and between it and the coast on both sides of the Medina every kind of marine industry is carried on, from yacht-building and the making of cordage and sails, more particularly for yachts, up to the building of 1600-ton torpedo boats and destroyers, merchant vessels of 4500 tons and 40,000 I.H.P., and every type of steam or oil-engine equipment.

Cowes is an interesting place or rather two places intermingled, the occupations of which dovetail one into another. The giant crane of White and Co.'s Shipbuilding Works dominates everything, towering above a mass of masts and cordage, elevators, warehouses, etc., and these combined with the crowded houses, and the narrow winding streets give a picture of great interest and variety.

Historic ground as Cowes is, as regards everything connected with yachting, it is also a very old shipbuilding centre, for the shipyard now carried on by Messrs Samuel White and Company was established in 1694, taking us back almost to Samuel Pepys and his official visits to Portsmouth. At that time the whole Solent area was busy with shipbuilding—Lymington, Buckler's Hard, Hamble, and Southampton Water all vied with Cowes and Fishbourne Creek, and many men-of-war were built and launched from this yard at Cowes. In 1885 the first 150-foot torpedo

Samuel White and Co.'s Shipyard and Engine Works, Cowes

"catcher"—they are now termed destroyers—built in this country was constructed here. The works cover 28 acres and give occupation on the average to 2500 men.

Elsewhere industries are quite minor. The milling of corn is still actively carried on. There is mention in Domesday of thirty-four island mills. The largest mills are at Newport, but on a number of the little streams, picturesque

Brighstone Mill Dam

mills are to be met with, as at Wootton Creek, Yafford, Alverstone, and Brighstone.

Brick-making is carried on at numerous places, and lime is burnt at Brading and elsewhere.

Fishing is only a minor industry. There is no regular fleet or fishing port, but a number of boats are owned at various parts of the coast, and lobsters and prawns are plentiful. Among the fish chiefly caught are whiting, plaice, a certain amount of bass, and mackerel. In the shallows along Newtown estuary oysters are cultivated.

The Newtown oysters are particularly large—larger than popular taste demands. The young oysters are brought from Holland and laid down in the river to fatten.

Of industries extinct or moribund the chief are stone-quarrying and the manufacture of salt and alum. The stone quarries at Binstead were once of extreme importance. The shell limestone of the Bembridge limestone formation was a splendid building stone used for Winchester and Beaulieu,

Bed of disused Saltern, Mouth of Newtown River

as also Quarr Abbey and many other churches both in the island and elsewhere. These special quarries are practically all worked out, and though stone is quarried elsewhere in the island the industry is quite small, and brick has taken the place of stone for modern building purposes.

Salt was obtained from "salterns," or saltpans, along the northern creeks. Those at Bowcombe, Watingwelle, and Whitwell are mentioned in Domesday. The industry died out when salt began to be extracted from brine, and an attempt was afterwards made to employ the disused salterns for oyster culture. For this purpose earthenware tiles were laid down in them to develop the "spat," but the salterns were in turn abandoned for the natural adjacent shallows. These dry beds now present a curious appearance from the tiles jumbled confusedly over them.

13. Shipping and Trade.

When one considers how the island waters are crowded with craft of every description, from 50,000-ton liners to tramp steamers, tugs and sailing craft of practically every nationality, passing and repassing daily along the Solent and Spithead, it is difficult to realise that practically none of this shipping, save yachts and the small packets and coasters passing between the island and mainland, has business with the island at all. It is the presence and situation of the island which has endowed the neighbouring ports and inland waters with their shipping facilities, but the island itself has no part in them.

The earliest ports of the island, Yarmouth (Eremue) and Newtown (Francheville), have long been surpassed by Ryde, Cowes, and Newport. Ryde is the packet station for Portsmouth and Southsea, Cowes for Southampton, and Newport for coasting vessels. Yarmouth still links up with Lymington, but except for ferry purposes it is entirely somnolescent. Excursion steamers carry large numbers of holiday-makers to and from other points where there are piers, as Sandown, Shanklin, Ventnor, etc., and Bembridge has from time to time been used as a packet station. Near Wootton Creek also a little coasting trade is done.

The one important form that maritime interests assume in the island is that of yachting. In this respect it gives the lead to the whole world. Yachting sums up Cowes and Cowes and the Royal Yacht Squadron almost sum up yachting. The three-branched character of the sheltered

Royal Yacht Squadron, Cowes

marine channels, their secure and easy anchorage in bad weather, and the breadth and depth of water are the chief advantages they possess. With Cowes at the centre a course can always be set either east or west according to conditions of wind and tide. There are numerous yacht clubs, but *the* club is of course the famous Royal Yacht Squadron. Vessels of the R.Y.S. above 30 tons have the unique privilege of flying the White Ensign and are admitted into foreign ports free of harbour dues. The Squadron headquarters are at the "Castle"—erected in 1856 on the site of the old blockhouse of Henry VIII's reign (West Cow), the semi-circular platform of which is now the only original portion remaining. Nowadays "Cowes Week" has been shorn of much of its glory, owing to the expense of keeping the larger types of yacht in commission, and only smaller racing craft have figured at Cowes since the war.

Many naval pageants, *e.g.* naval reviews, and the reception of the French fleet, and the visit of the late Czar and Czarina to England have taken place on these waters. Here the British Navy was mobilised in readiness for the great war of 1914—but these are rather national than island matters.

The island has a melancholy record in the matter of wrecks—among these we may record the wreck of the *Clarendon* in Chale Bay in 1836, and that of H.M.S. *Eurydice* off Dunnose in 1878, and going back a century or so it had an evil reputation for smuggling and for wrecking. Chale Bay, Puckaster Cove, and Luccombe were noted spots for running contraband goods. Blackgang had an unenviable notoriety also for "wrecking," and so serious an offence was this regarded that it was punishable with death.

14. History.

In following out island history one fact stands out clearly, namely, that apart from direct political events, the island has always flourished when its coast has been secure from attack—when this security has been menaced its prosperity has declined.

We will begin our story in A.D. 685, when, as previously related, Caedwalla of Wessex possessed himself of the island. Caedwalla at that time had himself only recently embraced Christianity and he at once proceeded to evangelise the island, which he did, as Bede tells us, by bestowing it for the service of God on Bishop Wilfrid. He made a grant to Wilfrid of one-quarter of the island, and the bishop sent two of his clergy to minister in the island—"one Bernwin his sister's son, and a priest named Hiddila." But the Channel was still in danger from pirates, and so a short time later, as Bede tells us, "no one was appointed to the ministry nor to the Bishop's seat until the time of Danihel who is now Bishop of Winchester." Thus in Bede's day the island had definitely become part of Wessex and included in the diocese of Winchester.

Before the Norman Conquest, however, the island had but little importance, being a sort of No Man's Land, continually menaced by attack from sea, rather than a possession to covet or desire. The Norman Conquest changed all that and gave the island a definite political position. Between that time and the present we can distinguish three periods of control.

(1) A period of almost independent rule under a feudal lord, during which no direct lordship was exercised over the islanders by the Crown. This lasted to A.D. 1293.

(2) A period when the Crown repossessed itself of direct lordship, acting either through wardens or through a Lord of the Island possessing hereditary rights; from 1293 to 1483.

(3) The subsequent period during which the Crown has acted through representative nominees—captains or captains and governors—not possessed of any territorial rights or independent position.

The advent of a Norman feudal lord gave a great impetus to island life. William FitzOsborn, Earl of Hereford, to whom William the Conqueror granted the island, marked his rule by building a big Norman keep at Carisbrooke, and made a grant of six island churches—Arreton, Freshwater, Godshill, Niton, Whippingham, and Carisbrooke to the Abbey of Lire, in Normandy. Thus a new element was brought into island life. Subsequently the lordship descended to the de Redvers family, which held it to A.D. 1293. One de Redvers granted a charter to Yarmouth (Eremue) which at this time was the port of the island, and later (in A.D. 1131) founded Quarr—the great Cistercian abbey which for 400 years was to be the chief civilising influence over the whole island.

The founding of Quarr brought importance to the east end of the island—and not merely Quarr alone, for the Binstead stone quarries had by this time become developed and the stone was in great request for building both in the island and on the mainland. The supremacy of Yarmouth as a port, however, was still unchallenged.

In A.D. 1160 (*circa*) another de Redvers granted a charter to Newport, and the Bishop of Winchester a similar charter to Newtown (Francheville). Somewhat later Brading began to develop. Thus we get a fairly definite picture of the island in the 12th century. A central seat and fortress, Carisbrooke, whence all influence emanated; three "Liberties"—the marks of growing importance—at Eremue, Francheville, and Newport; an abbey of commanding importance at Quarr; a thriving stone quarrying industry at Binstead—and continual backward and forward intercourse between island and mainland.

The last de Redvers to hold the lordship was Isabella de

Fortibus, famous for the resolution with which for many years she resisted the efforts of Edward I to wrest her rights from her. At last in 1293 on her deathbed she consented to sell them to the Crown, which was thus relieved from a position at once anomalous and dangerous to its authority.

From this date forward to Henry VIII's reign the island history is one of development checked from time to time by frequent foreign invasion. Edward III's wars brought insecurity, to guard against which Carisbrooke was strengthened, and defences were erected at Yarmouth, Shamblord (East Cowes) and La Rie (Ryde). The last named shows that the protection afforded by Portsmouth and the English fleet caused the eastern corner to grow more important. In spite of this, several ruinous French attacks were made—Yarmouth, Newtown, and Newport were all burnt, and Carisbrooke itself threatened. As a result land went out of cultivation and the population seriously declined. Newport, secure under the protection of the Castle, now began to forge ahead, and to outdistance Newtown, which never recovered.

With Henry VII's accession (A.D. 1485) the policy of appointing merely nominal "captains," instituted by his predecessor two years before, was confirmed. His successor's reign brought two heavy blows on the island—foreign attack, the result of anti-French policy, and the dissolution of Quarr.

In 1544 French invasion on a large scale threatened the whole southern sea-board. A French fleet sailed by Spithead, where it encountered the English fleet—the occasion when the famous episode occurred of the sinking of the *Mary Rose*, the King's flagship, with over 700 mariners, all drowned in the presence of the King. A French force actually landed on the south of the island and did considerable damage. To counteract such dangers the island defences were greatly strengthened. Worsley, the captain, erected stout blockhouses on both sides of the Medina mouth—the East Cow and the West Cow. He built a fort at Sandham (now Sandown), one at Yarmouth, and another at Yarmouth Common, opposite Hurst. He had previously erected forts or castles on the mainland at Hurst, Calshot, Netley, and Southsea, so that all the island waters were now strongly fortified. To supplement the shore defences each island parish was ordered to provide and maintain "a falconet of bronze and yron" and "gun chambers" in which these guns were kept were built on to the parish churches. The Brading gun is still preserved at Nunwell and the Carisbrooke gun in the Castle Museum.

The dissolution of Quarr was a calamity for the island. As a centre of piety and learning, no less than as one of skilful husbandry, it had dominated island social life just as the Castle dominated its political administration. When it was closed, all these beneficent activities ceased. The site and fabric were sold A.D. 1537 to a certain John Mylle, a merchant in Southampton.

Beyond a strengthening of the outworks of Carisbrooke Castle at the time of the Armada, and the building of a small fort, Carey's Sconce, near Yarmouth, later Tudor days brought little outward change to the island, but security from danger by sea and half a century of maritime adventure added enormously to its prosperity. In 1585 three towns—Yarmouth, Newtown, and Newport—sent members to Parliament, and Brading was later added to this list. The seaports at the eastern end grew in importance,

Quarr Abbey

and Cowes, Ryde (La Rie or La Riche), and St Helen's became increasingly active.

From this point onwards the island history relates to incidents rather than movements. Island security has never since been threatened. With the growing importance of Portsmouth and the Navy, Ryde and St Helen's came more and more into prominence, and from this time onward the western end has steadily diminished in importance and the eastern end and Cowes have as steadily advanced. All through the Dutch and French wars this process has continued, and the frequent references to the island in Marryat's books show how its progress has been involved in the growth of Spithead and of the Navy.

A leading incident of island history is the story of King Charles's imprisonment in Carisbrooke Castle. On Nov. 11, 1647, King Charles escaped from Hampton Court, and made for the Hampshire coast at Titchfield, with the hope that Col. Hammond, then Governor of the island, would befriend him. On being made a party to the King's plans, Hammond crossed over and escorted the King to Carisbrooke, where for nearly a year the King was kept a prisoner. He made several attempts to escape but they all proved abortive. In October, 1648, an effort was made to negotiate terms between the King and the Parliament; he was permitted to leave Carisbrooke and was lodged at Newport Grammar School, where for several weeks he held a so-called court. No result, however, was reached and on November 29th orders were received for him to be removed. He was accordingly escorted to Yarmouth whence he embarked for Hurst. Here he remained till December 18th, 1648, when he was taken to London for trial and subsequent execution.

A touching sequel of his connection with Carisbrooke was the death in 1651 of his daughter, the Princess Elizabeth. The chamber in Carisbrooke Castle in which she died is still shown, and many Stuart relics are preserved in the Castle.

Charles I

Appreciation of natural beauty, fostered by Wordsworth and the Lake poets, began to assert its influence in the 19th century. Sir Henry Englefield, in a classic work, described the beauties and interest of the island, and it sprang rapidly into popularity. Keats stayed here in 1817, writing part of *Endymion* at Carisbrooke and *Lamia* at Shanklin. George Morland, the painter, spent much time here. George Wilkes had a "villakin" at Sandham. Tennyson made Freshwater his residence, and the western end of the island is rich in memories of him. His lyric *Nightingales warbled without* is described in the title as having been written "in the Garden at Swainstone," the historic residence of the Simeon family, near Calbourne. Queen Victoria purchased the estate of Osborne, and Osborne House—built under the supervision of the Prince Consort—remained for many years her favourite residence. Cowes, Osborne, and Whippingham are specially connected with the Royal Family, and the Queen's youngest daughter, Princess Beatrice, is to-day Governor of the Isle of Wight.

15. Antiquities.

Compared with some other localities the antiquities of the Isle of Wight are not specially remarkable; but their study is of much interest and suggests many important lines of research.

Of early man, as already stated, the remains are scanty, and not very illustrative. It is only when the Bronze Age is reached that we come across many objects of interest. Of this period, however, the island can show a rich collection of "finds." The round barrows in which Bronze Age man buried his dead after incineration are found at many places over the heights; and associated with these are beautifully fashioned bronze implements, celts, spearheads and so forth. At Arreton and Brook palstaves, flanged celts, and knife daggers of extreme delicacy of workmanship have been found, and specimens of these are to be seen in the British Museum and in the island museum at Carisbrooke Castle. About thirty socketed bronze celts were disinterred at Ventnor while excavating for the Central Isle of Wight Railway. Bronze Age man seems to have been the first inhabitant of the island in the true sense of the word.

From this period on to Roman times there are many earthworks of uncertain date, some of the Bronze Age, some later. Black Barrow on Mottistone Down is the chief barrow, but numerous others exist, as at Ashey, Wroxall, Chilleston, etc. The series of depressions on Brighstone Down, marked on Ordnance Survey maps as British villages, are, however, now known not to be of human construction but of natural origin.

Of the Roman period of occupation, two survivals of outstanding interest exist, the Roman villas at Carisbrooke, and at Morton, near Brading. A smaller villa has also been discovered near Gurnard Bay, and some minor finds have occurred elsewhere. But in spite of many vigorous attempts to prove the contrary, no definite evidence of any Roman road has been found, and the attempt to build up from the name the existence of a Roman port at "Puckaster" Cove is dismissed by Prof. Haverfield as purely fanciful.

Of the Roman villas that near Brading is of remarkable interest and archaeological value. Discovered in 1880, it has been carefully and systematically excavated and the whole foundation of a wealthy Roman's country residence has been brought to light. A number of large and beautifully executed tessellated pavements, as well as the whole economy and internal arrangement of the group of associated buildings can here be studied. Much miscellaneous material, such as nails, tiles, pottery, coins, etc., has also been found and the whole is displayed to excellent advantage. The villa at Carisbrooke, discovered in 1859, is not so complete. It includes two large halls and some fine mosaic pavement. Considerable quantities of miscellaneous Roman finds have come to light from time to time at various other places.

The other island antiquities of outstanding importance belong to the Jutish and Early Saxon periods. These are the spoil of three places of interment on Arreton, Chessell, and Shalcombe Downs. The first discoveries were made in 1816, when important objects were brought to light, but Mr G. Hillier, working at Chessell in 1855, discovered 100 skeletons and a wonderful series of objects interred with them—shield bosses, swords, and other weapons of the men, brooches, buckles, jewellery, beads, and so forth of the women. These objects are exhibited in the British Museum, and no student of island antiquities should neglect to see them. Among them are two remarkable balls of crystal, about two inches and an inch and a half in diameter respectively, held in bronze frames and worn as pendants.

These interments belong to the early 6th century, the period corresponding approximately to our earliest extant references to island history, and their value as direct evidence as to racial origins in the Isle of Wight can be readily appreciated.

The "Gate House" Museum in Carisbrooke Castle, in addition to a good collection of pre-Roman antiquities, possesses a considerable number of Saxon and Jutish objects of the 6th and 7th centuries, and a large number of miscellaneous relics of island history.

Mediaeval and later life is reflected in the island by the bull ring and the ancient stocks and whipping-post preserved at Brading. The massive iron ring is seen let into the ground in the chief open space of the town halfway up the hill. The stocks and whipping-post tell of the days when to be a "masterless man" was equivalent to being a "rogue and vagabond."

The Stocks, Brading

Though they can scarcely be classed under the name of antiquities, we may here mention the numerous obelisks, beacons, and other landmarks erected on so many high points of the Downs and forming a striking feature in island landscapes, often of considerable service to the mariner. Ashey Beacon is a triangular pyramid on Ashey Down, and the Yarborough Obelisk, which commemorates the

Earl of Yarborough, founder of the R.Y.S., stands above Culver Cliff. Just behind Appuldurcombe is Worsley's Tower, a memorial of Sir Robert Worsley—the builder of the house—and Hoy's Monument, a lofty pillar at the end of St Catherine's Down records the visit of the Emperor Alexander of Russia to the island in 1814. The Tennyson Beacon, a tall cross in a commanding position on Freshwater Down, was erected in 1897 to commemorate the poet's long connection with the island.

16. Architecture—(*a*) Ecclesiastical.

A preliminary word on the various styles of English architecture is necessary before we consider the churches and other important buildings of our county.

Pre-Norman or, as it is usually, though with no great certainty termed, Saxon building in England was the work of early craftsmen with an imperfect knowledge of stone construction, who commonly used rough rubble walls, no buttresses, small semicircular or triangular arches, and square towers with what is termed "long-and-short work" at the quoins or corners. It survives almost solely in portions of small churches.

The Norman Conquest started a widespread building of massive churches and castles in the Continental style called Romanesque, which in England has got the name of "Norman." They had walls of great thickness, semicircular vaults, round-headed doors and windows, and massive square towers.

From 1150 to 1200 the building became lighter, the arches pointed, and there was perfected the science of vaulting, by which the weight is brought upon piers and buttresses. This method of building, the "Gothic," originated from the endeavour to cover the widest and loftiest areas with the greatest economy of stone. The first English Gothic, called "Early English," from about 1180 to 1250, is characterised by slender piers (commonly of marble), lofty pointed vaults, and long, narrow, lancet-headed windows. After 1250 the windows became broader, divided up, and ornamented by patterns of tracery, while in the vault the ribs were multiplied. The greatest elegance of English Gothic was reached from 1260 to 1290, at which date English sculpture was at its highest, and art in painting, coloured glass making, and general craftsmanship at its zenith.

About 1300 the structure of stone buildings began to be overlaid with ornament, the window tracery and vault ribs were of intricate patterns, the pinnacles and spires loaded with crocket and ornament. This latter style is known as "Decorated," and came to an end about 1350 with the Black Death, which stopped all building for a time.

With the changed conditions of life the type of building changed. With curious uniformity and quickness the style called "Perpendicular"—which is unknown abroad—developed after 1360 in all parts of England and lasted with scarcely any change up to 1520. As its name implies, it is characterised by the perpendicular arrangement of the tracery and panels on walls and in windows, and it is also distinguished by the flattened arches and the square arrangement of the mouldings over them, by the elaborate vault traceries (especially fan-vaulting), and by the use of flat roofs and towers without spires.

The mediaeval styles in England ended with the dissolution of the monasteries (1530–1540), for the Reformation checked the building of churches. There succeeded the building of manor houses, in which the style called "Tudor" arose—distinguished by flat-headed windows, level ceilings, and panelled rooms. The ornaments of classical style were introduced under the influences of Renaissance sculpture and distinguish the "Jacobean" style, so called after James I. About this time the professional architect arose. Hitherto, building had been entirely in the hands of the builder and the craftsman.

Apart from St Catherine's Tower and the ruins of Quarr the ecclesiastical buildings of the island are all parochial in type. Alien priories had been founded in the 12th century, but were all suppressed in 1414, and the only existing island church not strictly parochial is the Chapel of St Nicholas in Carisbrooke Castle, beautifully restored in recent days in memory of the unfortunate Charles I, so long imprisoned here.

Speaking broadly, the island churches are ancient. Many are Norman in origin, but all show considerable subsequent additions or alterations. None is specially large, but several are important-looking, and a number have really good towers.

Of church life before the Conquest we have no definite knowledge beyond the reference in Bede, already quoted, to Bishop Wilfrid and his priests Bernwin and Hiddila. There seems little reason to question the tradition of a 7th century church having been erected by Wilfrid at Brading, or to doubt that connecting St Boniface with Bonchurch and St Boniface Down, towering above it. For Winfrid, the Saxon monk of Nursling, it was who under the name of Boniface evangelised Germany, and as Archbishop of Mainz was martyred by the Frisians. We have no definite record of any pre-Conquest island churches, but the very name, Newchurch, applied to one of the oldest parishes is presumptive evidence of a pre-Norman church having existed here, nor would the Norman lord who erected Old Bonchurch church have dedicated it to a Saxon saint without a pre-existent similar dedication.

Norman Tower, Shalfleet Church

The Norman Conquest gave the first real impulse to island church building. We have seen how the first feudal lord, FitzOsborn, granted six churches—Arreton, Carisbrooke, Freshwater, Godshill, Niton, and Whippingham—to the Abbey of Lire, in Normandy. In addition to these, Domesday mentions two other churches—Bowcombe (*Bovecombe* or *Buccombe*) and Calbourne.

Of Norman work perhaps the finest example is the

tower of Shalfleet—a worthy example of 11th century Norman masonry, massive in form though of rather squat proportions. There is a good 11th century south doorway at Wootton church, with billet and chevron mouldings. Perhaps the most beautiful Norman work in the island is, however, the little church of Yaverland, where the original 12th century chancel, aisle, and south doorway remain. Binstead church—otherwise modern—still exhibits some original "herring-bone" work, sometimes claimed as pre-Conquest, and Bonchurch, Carisbrooke, Freshwater, Northwood, and Whitwell churches all show Norman features. Transition Norman (12th and early 13th century) is found at Arreton, Brading, Brighstone (the north arcade), Chale, Newchurch (both arcades) and Niton, as well as at Carisbrooke and Freshwater. Indeed, Transition Norman is the characteristic island type of church. Its wide distribution is evidence of great general development at this specially formative period of island history.

Of these Carisbrooke—considered to be the finest church in the island—has perhaps the most interesting history. The first church was early Norman, but later an alien priory was established here, and the church was reconstructed to serve both parish and priory at the same time. After the suppression of the priory, the property passed into the hands of Walsingham, Queen Elizabeth's minister, and he, as owner of the conventual portion of the church, pulled down the chancel and left the church in the disfigured state in which it is to-day. In spite of this, it is still a building of distinction, with a good timber roof, Transition arcade, and a beautiful Perpendicular tower. Arreton and Brading are both fine churches, partly Transition and partly Early English or later in character. The fine tower of Brading is a special feature and the Oglander chapel or mausoleum preserves the tombs and monuments of the great island family, the Oglanders of Nunwell. Churches with 14th century work include Arreton, Brading, Calbourne, Carisbrooke, Chale, Freshwater, Newchurch, Shalfleet, Whitwell, and Wootton.

The 15th century showed its influence by numerous additions being made to island churches—an instructive

Brading Church (*containing Oglander Mausoleum*)

evidence of activity especially in view of the heavy disasters which periodically afflicted the island about this time. This period is chiefly marked by the excellent towers at Carisbrooke, Chale, Gatcombe, Godshill, Shorwell, Whitwell, etc.

The east windows of the chancel and aisle in Calbourne church are quite unusual and have special interest, as they illustrate the development of the traceried window, from what is called plate tracery, *i.e.* a group of lights in combination, each pierced separately through the plain masonry of the wall, to the "regulation" single aperture, framed as it were in the masonry, and divided into separate lights by mullions. The Calbourne windows each consist of two lancet lights with a circular light placed centrally above, that in the chancel having a trefoil, and that in the aisle a quatrefoil.

Godshill

The chancel of Whitwell church reflects a curious little incident in local history. It was originally a chapel consisting of chancel and nave, and was built in the 12th century by the then lord of Gatcombe manor for his tenants at Whitwell. Later a second chancel and an aisle were added for the exclusive use of Godshill parishioners. A quarrel ensued between the two parishes, and after this had

Window in S. Transept, Calbourne Church
(Showing development of E.E. window)

been composed, the "wall of partition" between the two chancels was taken down, and the whole thrown into one.

Many of the churches, as having been built by lords of the manor for the benefit of their tenants, were originally very small. The old church of St Lawrence—until added to in 1830—had the distinction of being the smallest parish church in England. Kingston church, until 1766, retained its original dimensions—50 feet by 12 feet.

A feature of the island churches is the fine pulpits several of them contain. These naturally are mostly Jacobean, as *e.g.* the excellent panelled pulpit at Carisbrooke. The finest, however, is undoubtedly the beautifully carved Carolean pulpit of St Thomas, Newport, with its fine panelling and ornate sounding-board or "tester" of elaborate workmanship surmounting it.

The long tenure of estates by the great island families is shown by the monuments existing in many of the churches. A number of churches contain ancient brasses, and there are striking family monuments in many besides. They include those of the Oglanders at Brading, and the Leighs and Worsleys at Shorwell and Godshill, the Simeon chapel at Calbourne, the Afton chantry at Freshwater, and the Holmes monument at Yarmouth.

The island churches reflect in an especial degree the island characteristics, natural resources, and history. They are nearly all of stone—chiefly the Binstead limestone—though here and there, as at Mottistone, the ironstone, so abundant locally, is worked into the fabric.

17. Architecture—(*b*) Military.

Of military architecture in the island there is one outstanding example—Carisbrooke Castle. Its history sums up largely the political history of the Isle of Wight.

The site on which it stands is a commanding hill, and the presence of numerous mounds and early earthworks suggests that a fortress of some kind—a "burh" or stockade—existed here as early as Jutish or Saxon days. The Castle as we see it to-day was erected by William FitzOsborn, the first Norman feudal lord. It was repeatedly added to and strengthened in Norman days, and later during the early wars with France. The latest additions were made under Gianibelli, the famous Italian military engineer, in the days of the Armada.

In the reign of William and Mary it was allowed to fall into neglect. It now serves as the official residence of the Governor of the island—Princess Beatrice.

The general plan is that of a strong keep or donjon perched on the high ground at the north-east angle of an enclosure, surrounded by a battlemented *enceinte* of irregular oblong shape, with flanking towers at its angles known as Knights or Cavaliers. The outer or curtain wall surrounding it is of much greater extent and was the work of Gianibelli.

The outer gateway is Elizabethan and gives access to the Castle enclosure proper, surrounded by the mediaeval ramparts. The entrance to these is by the Gate House, erected in 1467 by Sir Antony Woodville, and called the Woodville Tower—a fine example of Edwardian military architecture.

The Keep is approached by a very long and lofty flight

of steps. The deep well within it was dug by an early feudal lord, Baldwin de Redvers, who being besieged in the Castle was forced to capitulate for want of water. On the Castle being restored to him later this well was constructed to guard against the repetition of a like mishap.

Behind the Keep is the *Place d'Armes* or exercising ground, converted into a bowling-green during King

Carisbrooke Castle

Charles's imprisonment here. The window from which he made his unsuccessful attempt to escape is shown, and also the room in which his daughter, the Princess Elizabeth, subsequently died. Of special interest is the chapel oi St Nicholas—the original castle chapel of 1070 now recently restored, as already related—and the famous well —not the one in the Keep already referred to—161 feet deep, from which water is raised by a wheel set in motion by a donkey. The important collection of island antiquities occupies the Gate House Museum.

Although several blockhouses—East and West Cow, Carey's Sconce at Yarmouth, and Yarmouth Castle—were erected at different spots around the island coast in the time of Henry VIII, nothing practically remains of them now except the base of the "West Cow," converted in 1856 into the Royal Yacht Squadron headquarters, and their importance in any case has never been great.

18. Architecture—(c) Domestic.

In domestic architecture the island has not only a real attractiveness, but special characteristics of its own. There are, of course, the usual commonplace modern dwellings, as in the outskirts of Cowes and the popular watering-places, but we find in the country areas inland or along the "back" of the island, a fine series of houses, of real character and distinction. The special Isle of Wight types are the old manor houses and the country cottages. In an area so abounding with building stone it would be singular if there were not a wide range of stone buildings, and such a range does exist, and admirable examples of it occur.

Real mediaeval domestic architecture is represented mainly by fragments. The older part of the Governor's Lodgings at Carisbrooke Castle dates from the 14th century. The great hall was built in 1386. Early English and Decorated work is to be seen in the traceried windows of Chale Manor Farm[1], and also in the remains of the bishop's

[1] These, incorrectly suggesting an ecclesiastical origin, have caused it to be misnamed Chale Abbey Farm.

palace at Swainstone. There is some doubt as to whether the remaining part of the Hall at Swainstone was ever used as a chapel. Stone's view is that it was used occasionally as an oratory, but not as a regular chapel. There are also Early English remains at the old house at Wolverton, near St Lawrence.

It is the fine series of Elizabethan and Jacobean manor-houses—often associated with characteristic stone cottages

Mottistone Manor Farm

of similar date, with mullioned windows—which are the real glory of Isle of Wight domestic architecture. These are in the main used as farm buildings now. The finest of these, the once famous house of Knighton, on Ashey Down, is unfortunately a thing of the past, having been pulled down in the 18th century. Mottistone Manor House (1567), disfigured by adjacent pigstyes, belonged originally to the Cheke family. Westcourt, near Shorwell, was mainly built in 1579, and Northcourt in 1615. Wolverton, also near

Wolverton Manor House

Shorwell, is of singular beauty (1600). So also are the manor-houses of Yaverland and Arreton—and there are many others. The large and indeed stately mansion of Appuldurcombe, the seat for many years of the Worsley family, dates from 1710. Many of the smaller houses, though not remarkable, have decided interest—they are shapely and have an air of distinction in spite of their limited size. Such, for instance, is the farm building at Walpen.

Buttressed stone barn, Chale

The Old Grammar School at Newport dates from 1619. It has a panelled oak room, used as a presence chamber by King Charles during the latter end of 1648, at the time of his temporary release from Carisbrooke. Newport Town Hall was built by Nash in 1814. Ryde Town Hall was erected in 1830.

The farm buildings throughout the island are often particularly attractive. The barns especially are of noble

Brick barn on "Staddles," Arreton Manor Farm

Ironstone in walls at Shorwell

proportions, well built, roomy, and at times imposing. The most striking perhaps is the great stone barn at Chale Manor Farm, with its fine open timber roof, and stone buttresses. Some of the barns are erected on staddles—even the brick-built ones—*e.g.* as at Arreton and Shalfleet, as a precaution against rats. The large size of the barns is explained by the dampness of the island climate, rendering out-of-door ricking undesirable. Modern days, however,

Half-timbered Cottage, Brading

are more utilitarian, and the modern farm building tends to be cheap, effective, and unsightly; hence corrugated iron is being increasingly used for sheds and outhouses.

Many of the old village cottages are of unusual charm, with generous overhanging thatched roofs, *e.g.* as at Godshill, Shorwell, and Brighstone. A few of these are still to be seen also at Shanklin Old Village, as it is termed.

The earlier buildings throughout the island are built

almost universally of island stone; in general either of Bembridge limestone or freestone, though here and there a considerable amount of ironstone is worked into them. There are a number of picturesque half-timbered cottages. In modern days brick has practically superseded stone.

19. Communications.

There is probably no feature of the island which has witnessed a greater transformation in the last hundred years than its roads and other communications.

The island trackways of pre-Roman days followed almost necessarily the crests or higher slopes of the chalk downs or other uplands, because these were then the only parts free from dense wood and scrub, and yet not rendered impassable by marsh and quagmire, as were the valley bottoms. These trackways exist to-day, and can be followed everywhere along the upper levels. There are no roads definitely established as "Roman roads" and for centuries communication was by mere natural tracks, over which people travelled on horseback. Wheeled traffic was till comparatively recent times all but unknown, and the crossing over from island to mainland was not only protracted and disagreeable, but was regarded even in Sir John Oglander's time as so dangerous that people regularly made their wills before embarking on the enterprise.

The first step towards improvement was as late as 1813, when a local Act was passed (53 Geo. III, c. 92), the preamble of which stated that "the public roads in the Isle of Wight are in many places in a very bad condition, narrow and incommodious, and in some places dangerous to travellers; and that they cannot be widened and repaired by the laws then in being" (*i.e.* by the old Highway Act of 13 Geo. III, c. 78, and by parochial surveyors under it). The new Act consolidated the parishes and placed the management under commissioners. Turnpike gates were set up and the island was divided into two districts, the East and West Medine, with a general road-surveyor appointed to each. The cost was apportioned to the various parishes. The beneficial results of this Act were seen in an immediate and marked improvement. Turnpikes were abolished in 1883. The present general road administration is now in the hands of the island County and District Councils.

Newport, the point to which island drainage in the main converges, serves also as the point of convergence of the roads. The roadways in fact form a general radial system, much like the spokes of a wheel, with Newport as the hub, while a practically continuous line of roadways more or less skirting the coast runs from Ryde through St Helen's, Bembridge, Sandown, Shanklin, Ventnor, the Undercliff, Chale, Brook, Freshwater, and Yarmouth, forming in part as it were the rim of the wheel. The northern coast fringe is impracticable for road traffic, and, accordingly, here the roads bend inwards from Yarmouth to West Cowes, and from East Cowes to Ryde.

Most of the roads are necessarily hilly, the two real exceptions being the so-called Forest Road from Parkhurst to Yarmouth, and the military road from Chale to Freshwater. The big chalk ridge from the Needles to Culver presents a number of depressions or gaps, which serve as "passes," over which the roads are carried. Much has been done quite recently to develop the road system. New roads have been cut, and others widened, and dangerous corners opened out.

In constructing some of the coast roads great difficulties of height and gradient have had to be overcome. The road from Sandown to Ventnor and Blackgang is a case in point. North of Shanklin it is practically at sea level, but to the south it mounts steadily to a height of 529 feet, sweeping round a fine curve at the back of Luccombe Bay, and commanding grand views seaward. Nearer Bonchurch it descends again rapidly with sharp "hairpin" bends. West of Ventnor the natural difficulties of a narrow fringe on the seaward side of the steep escarpment have been overcome by an elaborate series of zigzags. Beyond Chale the main road turns inland through Shorwell, Brighstone, and Mottistone, but a road known as the military road follows the line of the cliffs, and leads directly to Freshwater Gate. It was originally made about 1877, and, as its name implies, for military purposes. A good brick-built viaduct carries it over the mouth of Grange (or Brighstone) Chine.

It cannot be said that the island is well served with railways, in spite of their having been established under no less than five distinct companies, but many persons of course, with a wish to preserve for the island as far as possible its natural landscape beauties and retired character, feel quite content with the limited facilities at present existing. With the absorption of the separate Island railways into the Southern Railway, under the recent railway fusion scheme, improved facilities may be confidently looked for.

In the island, as elsewhere, development and ease of access have gone hand in hand. Ryde and Cowes, the nearest towns to the mainland, have been the first to develop. All along the stretch from Ryde and St Helen's to Sandown, Shanklin, and Ventnor development has been rapid and continuous. Yet there are other parts equally suitable, and it is expected that before very long they will be taken in hand, although undoubtedly one great check to progress is the necessity of transhipment caused by existing conditions of passage to and from the island, only to be overcome by a railway ferry or a tunnel.

As things go, however, the chief asset for island communication is the splendid pier at Ryde and the effective linking there of boat and train services. At Cowes the linking up is inadequate, and still more so on the Lymington and Yarmouth route.

Following the railway lines as they exist at present, it is interesting to notice that while the high roads generally avoid the bottoms of the river valleys, for reasons already indicated, this does not apply to the railways. The Eastern Yar curiously enough is followed more or less closely all along its course by one or other line of railway—its upper course by the line between St Lawrence and Merston; its next stretch by the Merston and Sandown Junction line, which crosses and recrosses it repeatedly; and its lowest reach by the railway which skirts the river all the way to Brading Harbour. A specially interesting point in the highway and railway problem is brought out along the Merston to Sandown section, viz. that while the highways all run directly across the valley at right angles to the contour lines, the railway follows the contour lines, *i.e.* the line of the stream. Hence railway and highway, which in ordinary valley conditions tend to follow parallel directions, are here found intersecting at right angles.

Water—except between Newport and Cowes—plays no

part whatever in the system of inland island communication; but as a medium for commerce the Medina estuary is a vital feature of transport within the island, a point already dealt with in a previous chapter.

20. Administration and Divisions.

At the time of the Domesday survey there were three Hundreds in the island, viz. the Hundreds of Bowcombe, Hamreswel, and Calbourne. Of these, Bowcombe Hundred occupied the major part of the island. Hamreswel Hundred, in which Yarmouth was included, disappeared at an early date, and by the 13th century, when the "Hundred Rolls" begin, there were only two Hundreds in the island, viz. the Liberty or Hundred of East Medine, and that of West Medine—the Medina stream forming practically the line of division.

Under ecclesiastical impulse the manors gave place, by a process of grouping, to a different unit for administration, viz. the "parish." Churches were built, often in the first instance, as we have already seen, for the spiritual benefit of the manor lord and his people. The process of formation of parishes was a gradual one, and their limits followed practically those of the manors of their founders. Hence the ancient parishes varied enormously in size, and their boundaries were often quite capriciously irregular. As population increased, and additional churches were needed, the larger parishes came to be divided into smaller ones, while more recently still in urban areas this process has been reversed, and "civil parishes" for rating and representative purposes have been formed by a re-grouping of contiguous ecclesiastical parishes.

A further variation in parish development took its rise about the 13th century. Religious houses had at that time acquired possession of many manors, and with them the corresponding tithes, and the practice arose of nominating a priest to carry out the spiritual duties of a parish, not as a "rector" or receiver of the tithe direct, but as a "vicar" or deputy, the tithe being paid to the monastery which thus itself remained the rector.

With the suppression of alien priories in 1414, and also when in Henry VIII's reign the monasteries were dissolved, the tithe as well as the monastic estates frequently came into private hands. We have already referred to an instance of this in connection with Carisbrooke (see *ante*, p. 97). In the island, the great abbey of Quarr, dissolved in 1536, owned tithes at Arreton, Haseby, Whitwell, Shalcombe, Luccombe, and Tidlingham, as well as the manors of Arreton, Newnham, and elsewhere, besides other island property. There were at this period no less than eight such island vicarages, whose "great tithes" were appropriated elsewhere—Arreton and St Nicholas Carisbrooke to the abbey of Quarr, and Carisbrooke, Brading, Newchurch, Godshill, Thorley, and Shalfleet to religious houses on the mainland. All these are now "vicarages."

Some of the ancient parishes were very large. The two largest, Newchurch and Carisbrooke, stretched from sea to sea across the island. Newchurch extended originally from Ryde to Ventnor, and with very few exceptions the ancient parishes all touched the sea at some point or other. Now, of course, many of the "ancient parishes" have been subdivided. Urban areas have grown into independent parishes, and these have later been further subdivided. Thus modern Ryde has grown up as a number of parishes, at the expense of the "ancient parishes" of Newchurch and St Helen's. Ventnor, in the same way, has been carved out of Newchurch, and the parishes of Sandown and Yaverland (the former recently, the last-named centuries ago) out of Brading. In like manner Northwood parish has grown out of Carisbrooke, and West Cowes out of Northwood. East Cowes is an offshoot of Whippingham.

With Plantagenet times came the growth of the "Liberties," *i.e.* areas (usually towns) with special privileges, granted them by charter, such as the right to hold markets, appoint their own officials, or bailiffs, as the functionaries now known as mayors were first called, etc. Yarmouth (Eremue) and Newtown (Francheville) have already been referred to in this connection. Newport received numerous charters, its first in the time of Henry II. In Edward I's reign a charter was granted to it by Isabella de Fortibus under the name of "her new borough of Medina," and it received later no less than fifteen Royal Charters. The present municipality exists in virtue of a charter of incorporation granted by James I, under which a mayor and burgesses were substituted for the bailiff. The oldest existing charter of Brading (*temp.* Edward VI) refers to earlier charters, and speaks of it as the "Kynge's town of Brading."

The present system of local municipal government dates from the Act of 1888, when the island became an administrative county, and the further Act of 1894, when urban and rural areas were formed, and placed under urban district councils and rural district councils respectively.

In the Isle of Wight there are two municipal boroughs,

Newport and Ryde, six urban districts, viz. Cowes, East Cowes, St Helen's, Sandown, Shanklin, and Ventnor, and one rural district, which comprises all the remainder.

For military purposes, until the passing of the Territorial and Reserve Forces Act of 1907, the island was under the sole control of the Captain or Governor, whose position in this respect has been practically that of Lord-Lieutenant. For Territorial Army purposes, in accordance with the Act of 1907, it is now under the Hampshire and Isle of Wight Territorial Army Association.

The administration of justice in early days was carried out by the lords of the island. The "Knighten Court" was held at Newport, under the steward of the lord, and had jurisdiction over the whole island, except the "Liberty of Newport." Later, the island became merged into mainland Hampshire for assize purposes, and this still remains, in spite of the formation of the island area into an administrative county. It is in the administration of justice only that the mainland county exercises any jurisdiction now over the island county. Newport and Ryde, as boroughs, have separate commissions of the peace—the remainder of the island forms one Petty Sessional Division. There are no island Quarter Sessions, and, as explained above, no island Assizes.

For Poor Law purposes, the whole island was formed into one Union. In 1771, the Isle of Wight Board of Guardians was first constituted. The "House of Industry," as the island "Workhouse" was named, is at Parkhurst, near Newport.

For Parliamentary purposes the island forms one constituency, and returns one member. At one time Yarmouth, Newtown, Newport and Brading all returned members.

The Reform Act of 1832 swept away all of these, except that of Newport, and Newport returned two members up to 1867, when its representation was reduced to one. The separate representation of Newport ceased in 1885.

Since Bishop Danihel's time, *i.e.* the 8th century, the island has been part of the see of Winchester. A scheme has been lately put forward officially for a tripartite division of the existing Winchester diocese, under which the Isle of Wight and the Portsmouth area would be formed into a separate see. No decision has at present been arrived at.

21. Roll of Honour.

The list of worthies connected with the island contains many historic names. Here we shall deal rather with the essentially island folk than with those greater historic personalities whose connection with the island was principally official or adventitious. Nor shall we refer specially to the feudal lords, or captains, though these often established a long family tenure, and their lines became essentially island families. Isabella de Fortibus, the last de Redvers to hold the lordship, Sir Richard Worsley of Appuldurcombe, "captain" from 1538 to 1565, Sir George Carey, "captain and governor" during the Armada period, are among the chief of these.

Prominent among island families of old and honourable standing are the Oglanders of Nunwell, the Worsleys of Appuldurcombe, the Leighs of Shorwell, the Simeons of Swainstone, and the Chekes of Mottistone. The Oglanders have held their estates since the Conquest, descending from

Richard de Oglandres, of Normandy, a follower of the Conqueror. The best known was the famous Sir John Oglander, the staunch loyalist and adherent of the Stuart cause, who died in 1655. He is best remembered by the valuable memoirs—the famous Oglander MS.—the classic source of information as to the island of his day. The Oglander tombs are in Brading church.

The Worsleys have had a long connection with the island. Appuldurcombe, originally an alien priory, was suppressed with others in 1414, but was ecclesiastically held for about a century, till it passed to Sir James Worsley in 1528 and was held in the family for three hundred years. His son, Sir Richard, succeeded to the captaincy and another Worsley built the present mansion. The estate passed from the Worsleys by marriage to the Pelhams, one of whom was the famous Earl of Yarborough, first commodore of the Royal Yacht Squadron.

The Simeons are an old family descended from Margaret Pole, Countess of Salisbury, beheaded in Henry VIII's time. The best known of the Chekes was Sir John Cheke, Regius Professor of Greek at Cambridge and Tutor of Edward VI, who, in Milton's words: "Taught Cambridge, and King Edward, Greek." Their seat was Mottistone, but Sir John Cheke was not born there.

Though Alfred Tennyson cannot claim a place in the Roll of Honour by right of birth, for he was born at Somersby in Lincolnshire in 1809, he is by residence, by his title, and indeed in almost every other respect, of the island. He came to Farringford in 1853, three years after succeeding Wordsworth as Poet Laureate, and spent the greater part of the remaining forty years of his life here and at Aldworth, hardly ever leaving his homes except for short journeys. He died October 6th, 1892, and lies buried in Westminster Abbey.

Dr Thomas James, the first librarian of the Bodleian Library at Oxford, born in Newport about 1570, is among the first of quite a long list of island worthies who attained great eminence in the University or academic world. Dr Thomas Pittes, an ardent Stuart adherent and a strong controversial divine, was another. He was expelled from Cambridge for his monarchical views and later became Chaplain to Charles II. Sir Thomas Hopson, afterwards admiral, served under Sir George Rooke at Cadiz and led the van at Vigo. He was a native of Bonchurch, and sat in Parliament later as member for Newtown. He died in 1717.

Robert Hooke, who as the author of "Hooke's Law" holds an outstanding position among English physicists, was born at Freshwater in 1635. He graduated at Christchurch, Oxford, and became Curator of experiments and later, for five years, Secretary to the Royal Society. Hooke's Law is the basis of the mathematical theory of Elasticity—stated in modern language it is "Strain is proportional to stress." Hooke published it originally in 1676 in anagram form ceiiinosssttuu, to which two years later he revealed the key—*ut tensio sic vis.* An irritable genius and bitter controversialist, he died in London in 1702.

Thomas Arnold—"Arnold of Rugby"—was born at East Cowes in 1795. His influence as Headmaster of Rugby, exerted not only in Rugby but on the whole public school life of England, is too well appreciated to need setting out here. Later he became Professor of History at Oxford. He died in 1842.

Several distinguished "overners" have held the living of

Brighstone. Three subsequently became Bishops—Ken, afterwards Bishop of Bath and Wells, one of the "seven bishops"; Samuel Wilberforce, Bishop successively of Oxford and of Winchester; and Bishop Moberly. Another clerical figure of note is Legh Richmond, whose name,

Thomas Arnold, D.D.

although his island residence only lasted eight years, is inseparably woven with island literature. Distinguished by his simplicity of life and genuine piety, this remarkable evangelical clergyman was curate of Brading with Yaverland from 1797 to 1805, and sprang into fame as the author of narratives of island life, chief of which were *The Dairyman's Daughter* and *Jane the Young Cottager*, published in 1814 in volume form as *The Annals of the Poor*. The memory of his heroines, Jane, buried at Brading, and Elizabeth Wallbridge, the dairyman's daughter, buried at Arreton, served for many years to attract visitors to the island from all over the world. Forgotten now, at one time they were the most widely read religious books in England, exceeding in popularity the *Pilgrim's Progress* itself.

Legh Richmond's books, apart from their very real spirit of piety and Christian faith, are of high literary level, and the descriptions of island scenery they contain have probably only been surpassed by one writer, Sir Henry Englefield. Though connected with the island only by a literary link, Sir Henry Englefield's magnificent work on island topography remains the classic work on this subject. It appeared in 1816 and is splendidly illustrated from his own sketches.

We have refrained of set purpose from any reference, beyond that in chap. 14, to Queen Victoria and members of the Royal Family, but memories of them are inseparably woven in the bright web and woof of island sentiment and cherished recollection.

22. Towns and Villages.

Arreton (932), a village 3 m. S.E. of Newport. Church with special features of interest to architects, and a good western tower. There is a fine Jacobean manor house adjoining it. (pp. 85, 91, 93, 98, 99, 100, 109, 111, 117, 124.)

Ashey (1471), civil parish adjoining Ryde. From Ashey Down adjoining, surmounted by Ashey Beacon, one of the best views in the island may be obtained. (pp. 92, 94, 107.)

Bembridge (1428), attractive seaside resort and yachting centre at extreme east of island. The Warner Lightship and St Helen's Fort lie in front of it and Bembridge Down and Whitecliff Bay adjoin it ($2\frac{1}{2}$ m.). The so-called Centurion's Copse adjoining is a misnomer, derived from St Urian, a Breton priest, to whom a small chapel formerly here was dedicated. (pp. 15, 31, 81, 112, 113.)

Binstead (969), village, originally ancient, now practically a suburb of Ryde. The site of famous quarries, where Bembridge shell limestone, the fine island building stone, was formerly worked. Adjoining it are ruins of Quarr Abbey, and a splendid modern building with

Arreton Manor Farm

fine church also called Quarr, erected by the Benedictine Order. The monks recently left here to return to Solesmes. (pp. 39, 41, 80, 85, 87, 97, 99, 103.)

Blackgang, hamlet and coastguard station at head of the famous Blackgang Chine. Above it towers St Catherine's Down, with the old tower or pharos. (pp. 13, 32, 43, 46, 50, 83, 114.)

Bonchurch (530), a village adjoining Ventnor, anciently Bonecerce or Boniface's Church, the traditional centre of the ministrations of St Boniface (Winfrid of Nursling). The ancient Norman church,

Brading Market-place and "Bull-ring"

in bad repair, replaced by a modern one, still stands. In the churchyard the poet Algernon Charles Swinburne is buried. (pp. 6, 32, 98, 99, 114, 122.)

Brading (1563). Formerly a municipal borough—Ye Kynge's towne of Brading. Has ancient charters and formerly sent two members to Parliament. Traditional scene of Bishop Wilfrid's labours. Fine church, Tr. Norman and E.E., with Oglander monuments. Nunwell, the seat of the Oglanders, adjoins it, and the fine Roman villa is at Morton

Brighstone, Characteristic Village Corner

hard by. The bull-ring, stocks, and whipping-post are still preserved. (pp. 20, 40, 50, 58. 79, 85, 87, 92, 94, 97, 99, 100, 103, 115, 117, 118, 119, 121, 123, 124.)

Brighstone, also spelt Brixton (469). Picturesque secluded village near the S.W. coast of the island. Fine Norman church, much restored. Ken, Wilberforce, and Moberly, afterwards bishops, were all rectors here. The manor was conferred by Egbert on the see of Winchester, whence the name (=Ecbright's Town). Grange Chine adjoins the village. (pp. 9, 79, 92, 99, 111, 114, 123.)

Calbourne (720), highly picturesque village, with church mentioned in Domesday: Norman with illustrative E.E. The plate tracery of east windows of chancel and aisle specially remarkable. Simeon chapel contains monuments of the family. Near by are Westover and Swainstone, famous old island seats, the latter of which belonged to the bishops of Winchester. (pp. 91, 98, 100, 101, 103, 116.)

Carisbrooke (5139), now practically a suburb of Newport, which grew round it under the protection of the Castle. Famous castle already described, as also church. The latter (Norman and later) has fine Perpendicular tower. The chancel pulled down by Walsingham

Carisbrooke, "Water-splash" over the Lukely

has never been replaced. A Roman villa was found here 1859. (pp. 9, 18, 40, 66, 85, 86, 87, 89, 90, 91, 92, 93, 97, 98, 99, 100, 103, 104, 106, 109, 117, 118.)

Chale (565), village lying N.W. of St Catherine's Hill, 230 feet above Chale Bay. Tr. Norman church with good Perpendicular tower. Chale Farm, adjoining, has interesting ancient buildings and barn. (pp. 15, 40, 46, 48, 50, 61, 62, 83, 99, 100, 106, 111, 113, 114.)

Cowes (Urban District) (9998), and E. Cowes (Urban District) (4636), the twin urban areas at mouth of Medina. Centre for yachting and marine engineering. The Royal Yacht Squadron and Cowes Week are famous. Has no ancient churches. The name derived from block-houses, West Cow and East Cow, built here in Henry VIII's reign. (pp. 6, 15, 20, 53, 54, 58, 62, 69, 76, 77, 81, 82, 83, 86, 87, 89, 91, 106, 113, 114, 115, 118, 119, 122.)

Wheel of Alverstone Mill

Freshwater (3192), parish with scattered houses. Ancient church of Norman origin, with Afton chantry, and Tennyson family memorials. The poet lived at Farringford, close by. (pp. 9, 14, 15, 21, 33, 41, 43, 45, 49, 55, 57, 85, 91, 95, 98, 99, 100, 103, 113, 114, 122.)

Gatcombe (369), small inland parish, 2 m. S. of Carisbrooke. Church (13th century) has a good tower, and contains curious effigies. (pp. 100, 101.)

Godshill (946), picturesque village 6 m. S. of Newport with charming thatched cottages. Church has a fine Perpendicular tower and monuments of Leigh and Worsley families: adjoining is Appuldurcombe House, the former seat of the Worsleys. (pp. 20, 85, 98, 100, 101, 103, 111, 117.)

Mottistone (100), now united with Shorwell. Church mainly Perpendicular, much restored. Above it on Mottistone Down is the well-known Longstone. There is a famous manor-house, formerly the seat of the Cheke family. (pp. 9, 64, 92, 103, 107, 114, 120, 121.)

Old Town Hall at Newtown

Newchurch (751), formerly the most extensive of ancient island parishes. Present church Tr. Norman, with rose window at west gable. (pp. 20, 98, 99, 100, 117, 118.)

Newport (11,036), ancient municipal borough with mayor and corporation. Developed under protection of Carisbrooke Castle as the centre of island government and trade, being well situated at the head of Medina estuary. Roads radiate from it to all parts of the island. Brewing, milling and timber industries are carried on, and it has the only stock-market in the island. At the Grammar School (1619) King Charles held his so-called court, 1648. Modern church contains fine woodwork, magnificent Carolean pulpit, and recumbent effigy of Princess Elizabeth. (pp. 10, 15, 18, 32, 42, 69, 72, 76, 77, 79, 81, 82, 85, 87, 89, 103, 109, 113, 115, 119, 120, 122.)

Newtown, formerly Francheville, was once a borough and sent two members to Parliament. Prosperity began to depart after the French burnt it in 1377. Nothing now remains but a few cottages, a church recently rebuilt, the coastguard station, and the old Town Hall, now practically ruinous. The ancient corporation regalia are preserved at Swainstone. (pp. 38, 50, 79, 80, 81, 85, 86, 87, 118, 119, 122.)

Niton (866), 1 m. from coast, adjoins the Undercliff. A delightful spot. Church Tr. Norman, with square tower and small spire. St Catherine's Lighthouse on cliff just above St Catherine's Point, about 1½ m. away. (pp. 40, 85, 98, 99.)

Northwood (2016), midway between Cowes and Newport, has a Tr. Norman church. Parkhurst Forest lies to the east, and Parkhurst Prison, the Barracks, and the House of Industry (workhouse) lie between it and Newport proper. (pp. 99, 118.)

Ryde (11,295), thriving municipal borough, originally La Rie or La Riche, only began to grow at end of 18th century. There are several churches, all modern, and a specially fine pier. It is the port for packet services to and from the mainland. (pp. 15, 54, 69, 76, 81, 86, 89, 109, 113, 114, 115, 117, 118, 119.)

St Helen's (5706) is a parish and urban district rapidly developing. The village round which it centres is small. The present church (1717) replaces the old church, which had to be pulled down owing to coastal erosion. The tower of the latter still stands as a beacon for mariners. (pp. 36, 37, 58, 89, 113, 115, 118, 119.)

St Lawrence (366), on Undercliff 2 m. E. of Ventnor. Ancient church, said to be the smallest parish church in England, until added to in 1830. (pp. 45, 62, 103, 107, 115.)

Sandown (7664), a rapidly growing seaside town in the centre of Sandown Bay, originally Sandham. Has a good pier and fine sands, and is renowned for its bathing. (pp. 14, 15, 20, 46, 49, 50, 54, 62, 69, 72, 81, 87, 91, 113, 114, 115, 118, 119.)

Seaview (971), a recently created ecclesiastical (but not a civil) parish, has risen into popularity as a centre for sea-bathing and other attractions. (p. 42.)

Shalfleet (822), inland village on road between Newport and Yarmouth. Church with fine Norman tower, and early Jacobean pulpit. (pp. 99, 100, 111, 117.)

Shanklin (7374), a rapidly growing town and favourite summer resort 2 m. N. of Ventnor. The "old village" has picturesque thatched houses, and a fountain with an inscription written by Longfellow. The famous Shanklin Chine is charming rather than impressive. (pp. 13, 15, 40, 49, 62, 69, 81, 91, 111, 113, 114, 115, 119.)

Shide and Blackwater are hamlets with railway stations in the Medina valley. At Shide Prof. John Milne established his famous seismological observatory. (p. 18.)

Shorwell (541), an inland village with charming thatched cottages, 5 m. S.S.W. of Newport. Church mainly Perpendicular, with a good tower, monuments of the Leigh family, and a St Christopher fresco. The parish is now united for church purposes with Kingston and Mottistone. The attractive manor houses of Wolverton and Northcourt are near. (pp. 42, 100, 103, 107, 111, 114, 120.)

Totland (1441) is rapidly developing as a place of resort in summer. Colwell Bay, Headon Hill, and Alum Bay adjoin it. (pp. 31, 43.)

Ventnor (6063), a quite modern town, in great favour as a winter resort. Built on a steep slope, its "hairpin bend" streets and roads have been ingeniously engineered, and above it towers the fine St Boniface Down. Ventnor has long enjoyed reputation as a health resort, and a little west of it is the large sanatorium for tubercular patients. (pp. 6, 15, 46, 69, 76, 81, 92, 113, 114, 115, 117, 118, 119.)

Whippingham (2545), the mother parish of East Cowes, is famous for Whippingham church (modern) and its associations with the Royal Family. Osborne House, Queen Victoria's favourite residence, was built here in 1851 and later made a convalescent home for officers. The Royal Naval College near by was closed in 1921. At his model farm at Barton in the parish, the Prince Consort did pioneer work, much aiding the island agriculture of his day. (pp. 85, 91, 98, 118.)

Whitwell (681), village 1½ m. N. of St Lawrence, on the road from Newport, with a church of the 12th century. (pp. 20, 80, 99, 100, 101, 117.)

Wootton (143), an ecclesiastical but not a civil parish, stands at the head of Wootton Creek. A small coasting trade is carried on here and formerly a considerable shipbuilding industry existed along the creek. The church is ancient, with a fine Norman south doorway. (pp. 50, 79, 82, 99, 100.)

Wroxall (856), pleasant village nestling in the downs about 2 m. N.W. of Ventnor. Appuldurcombe Park adjoins it to the north. There is much slipped gault in the neighbourhood. Between it and Shanklin are the beautiful America Woods. (pp. 20, 92.)

Yarmouth (847), at the mouth of the Eastern Yar, is a quiet little port, with a considerable marsh at the back of it at a very low level. In early days it was the principal port, and the earliest island borough (Eremue). It has a most interesting history. (pp. 15, 38, 50, 53, 54, 57, 62, 81, 85, 86, 87, 89, 103, 106, 113, 115, 116, 118, 119.)

Yaverland (135), a quite small parish adjoining Brading, with a Norman church, and a highly picturesque manor-house. (pp. 9, 99, 109, 118, 123.)

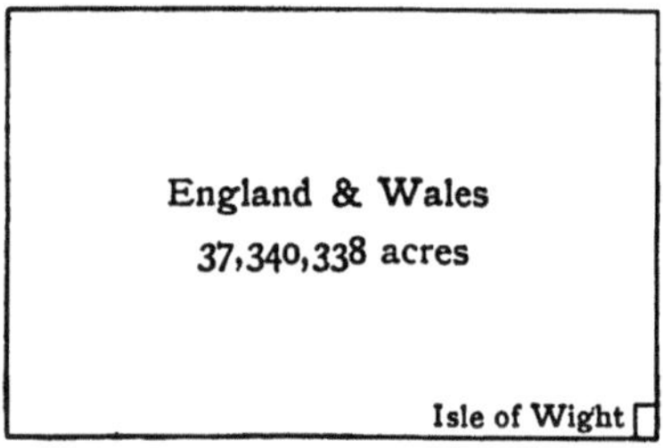

England & Wales
37,340,338 acres
Isle of Wight

Fig. 1. Area of the Isle of Wight (94,146 acres) compared with that of England and Wales

England & Wales
37,885,242
Isle of Wight

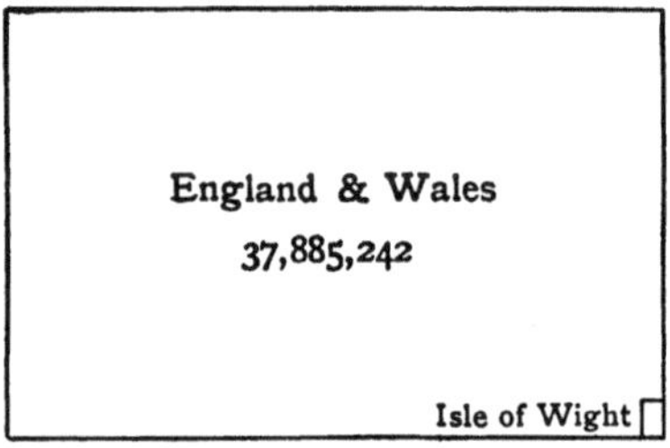

Fig. 2. Population of the Isle of Wight (94,697) compared with that of England and Wales in 1921

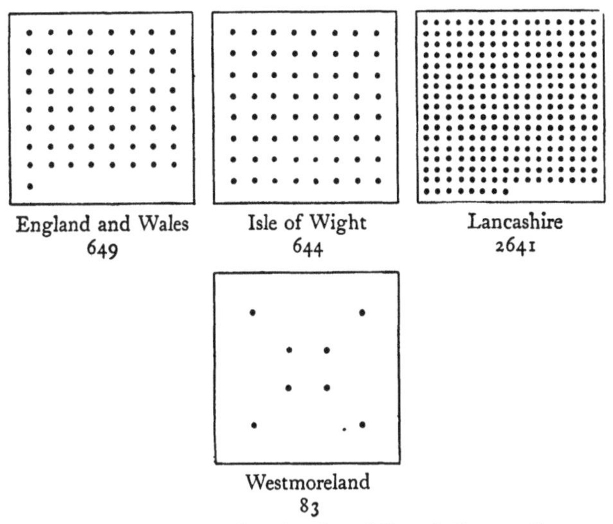

Fig. 3. Comparative density of Population to the square mile in 1921. (*Each dot represents ten persons*)

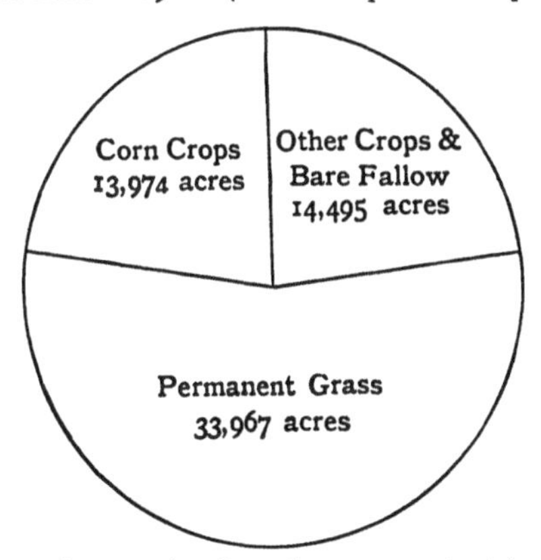

Fig. 4. Area under Cereals compared with other crops and grass in the Isle of Wight in 1922

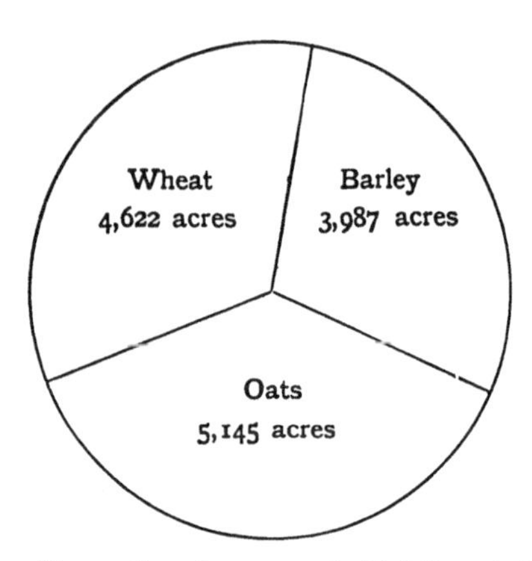

Fig. 5. Proportionate areas of chief Cereals in the Isle of Wight in 1922

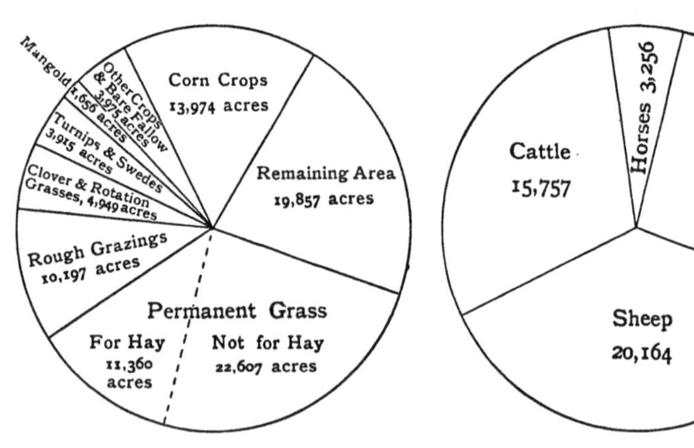

Fig. 6. Proportionate uses of land in the Isle of Wight in 1922

Fig. 7. Proportionate numbers of Live Stock in the Isle of Wight in 1922

www.ingramcontent.com/pod-product-compliance
Ingram Content Group UK Ltd.
Pitfield, Milton Keynes, MK11 3LW, UK
UKHW042146280225
455719UK00001B/129

9 781107 628700